Technician Class

FCC License Preparation for Element 2 Technician Class Theory

by
Gordon West
WB6NOA

with Technical Editor
Eric P. Nichols
KL7AJ

Tenth Edition

Master Publishing, Inc.

Also by Gordon West, WB6NOA
&
Eric P. Nichols KL7AJ

General Class
FCC License Preparation for
Element 3 General Class Theory
Also available as an audio book
and with W5YI Ham Study software

Extra Class
FCC License Preparation for
Element 4 Extra Class Theory
Also available as an audio book
and with W5YI Ham Study software

GROL + RADAR
General Radiotelephone Operator License
Plus Ship Radar Endorsement
FCC Commercial Radio License Preparation for
Element 1, Element 3, and Element 8 Question Pools
Also available with
W5YI Comm Study Software

About the Authors

Gordon West, WB6NOA

Gordon West has been a ham radio operator for more than 60 years, holding the top Extra Class license, call sign WB6NOA. He also holds the highest FCC commercial operator license, the First Class General Radiotelephone Certificate with Radar Endorsement. Gordon teaches evening ham radio classes and offers weekend ham radio licensing seminars on a monthly schedule. These seminars cover entry-level and upgrade licenses in ham radio. He has served on the faculty of Coastline College and Orange Coast College. Gordon is a regular contributor to amateur radio, marine, and general two-way radio magazines. He is a fellow of the Radio Club of America, and a life member of the American Radio Relay League. The ARRL presented Gordon with its "Instructor of the Year" award. The Dayton Amateur Radio Association named Gordon their "2006 Amateur of the Year" for his efforts in recruiting and training new amateurs, in addition to his lifelong involvement in ham radio. Through his Gordon West Radio School, he has trained eight out of ten newly-licensed hams with his classes, books and audio courses over the past 40 years.

About the Authors

Eric P. Nichols, KL7AJ

Joining author Gordon West for this new Tenth Edition of his Technician Class study manual is Eric P. Nichols, KL7AJ. Eric has been a licensed amateur operator since 1972, holding the top Extra Class license, as well as the FCC GROL commercial license. He has spent his adult career in various aspects of radio research, broadcast communications, and teaching. Eric is the proprietor of AlasKit Educational & Scientific Resources (AlasKit.net), which provides components and kits for experimentally minded radio amateurs, with a focus on our newest amateur radio bands on 2200 meters and 630 meters.

A prolific author, he has written many articles for ***QST, QEX,*** and a number of other ham radio magazines, as well as professional journals. In 2010 and again in 2014 he was the recipient of the William Orr, W6SAI, Technical Writing Award conferred by the American Radio Relay League for articles he authored for ***QST*** magazine. He is the author of ***Receiving Antennas for the Radio Amateur***, ***Radio Science for the Radio Amateur***, and ***The Opus of Amateur Radio Knowledge and Lore***.

Eric lives in North Pole, AK. He has operated just about every mode available to the radio amateur, but always returns to CW on the lower HF bands as his "default" mode. He enjoys restoring vintage "boat anchor" radio equipment as well as designing cutting edge radio instrumentation with Arduino, his latest obsession. Being in interior Alaska, he gets a lot of requests from new hams (and some older ones) for a first-time Alaskan radio contact, and is always willing to try a "sked" on any HF band or mode!

This book was developed and published by:
Master Publishing, Inc.
Glencoe, Illinois

Editing by:
Pete Trotter, KB9SMG, and Julie Scheife, Mayfly Design
Book typesetting by Mayfly Design

Photograph Credit:
All photographs that do not have a source identification are either from the author, or Master Publishing, Inc. originals.

Cartoons by:
Jim Massara, N2EST, based on the original Elmer by Carson Haring, AC0BU.

Thanks to the following for their assistance with this book: Suzy West, N6GLF; Dave W6DVE & Tina W6TNA Bremer; Jim Ford, N6JF; Chip K7JA & Janet KL7MF Margelli; Tracy Lenocker, WM6T; Jodi Lenocker, WA6JL; Rosalie White, K1STO; Stephen Gallagher, KE8TEY; Andrew Zuffi; Herbert Nigg; Lyle Long N1LL; Tim Sturm, N5OXY; Joe Leroux, KK7ASM; Randy Hall, K7AGE, and the many other hams with book suggestions! Thank you all!!

Printing by:
Arby Graphic Service
Glencoe, Illinois

Visit Master Publishing
on the Internet at:
www.masterpublishing.com

Tenth Edition
5 4 3 2 1

Table of Contents

QUESTION POOL NOMENCLATURE

The latest nomenclature changes and question pool numbering system recommended by the Volunteer Examiner Coordinator's Question Pool Committee (QPC) for question pools have been incorporated in this book. The Technician Class (Element 2) question pool has been rewritten at the middle-school reading level. This question pool is valid from July 1, 2022 until June 30, 2026.

FCC RULES, REGULATIONS AND POLICIES

The NCVEC QPC releases revised question pools on a regular cycle, and deletions as necessary. The FCC releases changes to FCC rules, regulations and policies as they are implemented. This book includes the most recent information released by the FCC at the time this copy was printed.

Preface

Welcome to the fabulous, fun hobby of Amateur Radio! It has never been easier to enter the amateur radio service than now!

Changes to Amateur Radio regulations by the Federal Communications Commission, which became effective April 15, 2000, have made it easier than ever to obtain your entry-level Technician Class ham radio license — and to move up through all the license classes to earn the top Amateur Extra Class license.

And when we say, "It has never been easier...," we really mean it! In years past, to achieve Technician Class level, you would have studied over 900 Q & As. Now, beginning July 1, 2022, we are down to 411 questions in the pool, and just 35 on the exam. And to better encourage youth participation in Amateur Radio, the writing is at an accessible middle-school reading level. Absolutely no knowledge of Morse code is required for this entry-level license, and math skills are limited to simple division and multiplication.

In just one exam session, you can satisfy all of the requirements to earn your Technician Class amateur radio license. Within days of passing your written exam (you need to answer 26 of the 35 questions correctly — that's just 74%) you can be on the air talking through repeaters, transmitting via satellites to work other stations thousands of miles away, and even getting a taste of some ionospheric skywave worldwide contacts, plus ham HD TV and digital modes!

On February 23, 2007, a new Federal Communications Commission ruling eliminated the Morse code test requirement for all classes of amateur radio licenses. The same ruling also granted all Technician Class operators limited portions and modes on four worldwide, high-frequency ham bands! So if the dots and dashes have kept you out of ham radio before, there is absolutely no code test required for Technician, General, or Extra class licenses.

Most important, after you pass the Technician Class exam, GET ON THE AIR! Throughout this book, Eric, Josh, and I give you lots of pointers to teach you how to be a good ham operator.

Ready to get started? Hurry up – I am regularly on the airwaves, and I hope to make contact with you very soon with your new Technician Class call sign!

Gordon West, WB6NOA　　　*Eric P. Nichols, KL7AJ*

About This Book

This book provides you with all of the study materials you need to prepare yourself to take and pass the Element 2 written examination to obtain your Technician Class amateur radio license. Technician Class is the entry-level amateur operator/primary station license issued by the Federal Communications Commission — the FCC. *Absolutely no Morse code test is required* for the Technician Class license, which will give you unlimited VHF and UHF ham band privileges, plus the excitement of operating on worldwide skywave bands, too!

Our book also provides you with valuable information you need to be an active participant in the amateur ranks. To help you get the most out of *Technician Class,* here's a look at how our book is organized:

- *Chapter 1* provides an overview of the amateur service and a quick look at all the exciting things you can do with your entry-level Technician Class license.
- *Chapter 2* tells you about all of the ham radio operating privileges you will have with your new Technician Class license — including the additional high frequency privileges you will earn when you pass your Technician Class exam.
- *Chapter 3* gives an overview of the amateur service, and a brief history of ham radio regulations. It contains details on the 1999 FCC Report & Order that greatly simplified the Amateur Radio service licensing structure, which streamlined the number of examination elements and reduced the emphasis on Morse code for all classes of ham radio licenses. It's your orientation to ham radio.
- *Chapter 4* describes the Element 2, Technician Class written examination, and contains all 411 middle-school-reading-level questions that comprise the Element 2 question pool. Thirty-five of these multiple choice questions will be on your written examination. If you answer 26 of them correctly (74 percent), you will pass the exam and receive your FCC license. Our book and companion audio course relate every question in the pool to the real world of operating and getting on the air with ham radio! And to further your learning, be sure to check out the Ham Radio Crash Course videos noted at the end of each topic section.

Watch This! **https://www.youtube.com/playlist?list=PL1KAjn5rGhizk5f2_whBLJaOHj59OCI_3**

- *Chapter 5* will tell you what to expect when you take your Element 2 written exam, where to find an exam session, how you will apply for and receive your new FCC license, and more — all the details you need to know to get your license and get on the air.
- *Chapter 6* talks about Morse code. Even though you don't need the Morse code to become a ham operator, learning code with our audio course is fun, and all new ham operators should learn the code as a basic worldwide language of dots and dashes that will live on forever – especially in emergencies.
- *The Appendix* has valuable lists and reference information. And reading the *Glossary* will get you up to speed on some of the amateur radio "lingo" that you might not understand as you begin studying our book.

1

Getting Into Ham Radio

WELCOME TO AMATEUR RADIO!

There are many great stories about the origin of our "ham" nickname, and soon you will decide on your favorite. Many people call us "hams" because we always seem to be ready to "show off" the magic of our little wireless gadgets. It won't be long before you'll be doing this, too!

Amateur radio operators INVENTED wireless communications. Did you know that Guglielmo Marconi – who often is credited with inventing radio – considered himself an amateur? Amateur operators were 120 years ahead of your little cell phone. Our century-old Morse code dots and dashes were a forerunner to that little micro tablet that's tucked away in your pocket or purse. Just ask any ham operator who knows his history, and you'll be told that this new hobby and service you are preparing to join truly did shape everything that is going on with wireless communications today and in the future.

There's a lot of fun to be had in the ham radio hobby you're about to join!

Nearly every country in the world has an amateur radio service, and we all share certain ham radio frequency bands globally. There are more than 5,000,000 licensed amateur radio operators throughout the world, and we number more than 842,000 here in the U.S.

WHY DO I NEED A LICENSE?

Unlike a lot of electronic communications devices that you already use – like cordless phones in your home, cell phones on the road, or short-range FRS or CB radios – the ham radio equipment you will learn to use has capabilities to communicate across town, around the world, and even into outer space. So, in order to keep things orderly, all hams are required to demonstrate that they understand the rules, regulations, and international frequency assignments placed on amateur radio.

All countries require their ham radio operators to be licensed; to know their country's local radio rules and regulations, and to know a little about how radios work in order to pass that "entrance exam." Here in the U.S., the exam for the entry-level Technician Class amateur operator license is a snap. Chapter 4 gives you all the exact exam questions word for word!

Once you pass the 35-question multiple choice "entrance exam," you'll be issued your first amateur radio license by the FCC (Federal Communications Commission). Absolutely no knowledge of Morse code is required for your Technician Class radio license, or any FCC ham license.

Your brand new Technician Class license will authorize you to operate with unrestricted access on all ham bands above 50 MHz, including exciting voice privileges on the long range 10 meter band, plus 3 other worldwide bands for Morse code! Ham "bands" are internationally-designated groups of frequencies reserved for amateur radio operation. In the next Chapter, and throughout this book, you'll learn about frequencies in cycles per second (hertz), ham bands in wavelengths, and exactly where our ham bands are located on the radio dial.

YOUR FIRST RADIO

The first radio that most Technician Class operators start out with is a dual-band handheld transmitter/receiver. You'll tune into frequencies that are automatically relayed to other ham radio operators throughout your area via repeaters. The keypad on the front of your little handheld may also be used to dial up internet radio links. Imagine walking down the street with your handheld talking to a fellow ham radio operator in Australia – or maybe in Antarctica! Or how about calling home on a free ham radio "autopatch" network – or listening to freeway traffic reports that are more frequent and accurate than those on your car radio!

You'll probably want to get a dual-band handheld transceiver for your first ham radio.

Many dual-band handheld ham radios also have the capability to tune in worldwide shortwave broadcasts, AM and FM radio stations, fire frequencies, and even do a little eavesdropping backstage on wireless microphone frequencies. These are very sophisticated pieces of radio equipment with many fun and interesting capabilities designed in.

Later, after you've learned how to operate properly and are ready to expand into other ham radio capabilities, you can consider buying a mobile radio that can be mounted in your car or pickup, and even used as a "base station" in your home. But our strong recommendation for your first radio is the dual-band handheld.

After you've developed your operating skills, you may want a mobile set for your car or pickup. They can be used as base stations in your house, too.

What about a radio tower? Well, while the Eiffel Tower in Paris was one of the world's first antenna towers, today's antennas for ham radio can be very discreet. They can be safely placed in attics, or be stealth hookups to a nearby window screen. So, no, you won't need a huge tower to get started in your new hobby and service.

WHAT ELSE CAN YOU DO? PLENTY!

Your new handheld can plug into a global positioning system (GPS) and relay your location to ham satellites or mountaintop digital repeaters tied into the internet. Your fellow hams – even mom – could track your every movement. Or maybe you'd like to plug your handheld into a tiny color video sender and show all the hams around you what it looks like from the top of that mountain you just climbed!

Your new Technician Class license allows you every ham radio privilege on all of the VHF and UHF bands. Imagine talking thousands of miles away by bouncing 6-meter band signals off of the ionosphere. Six-meter skywave excitement occurs summer and fall, and many Technician operators have worked hundreds of other ham stations throughout the world on 6-meter single sideband. You also gain long-range skywave privileges on a portion of 10-, 15- 40-, and 80-meters, too!

Are you joining us from Citizens Band radio? If so, your 27 MHz antenna can easily be adjusted for 10 meter, 28 MHz, worldwide voice excitement!

What else can you do with all of your new Technician Class privileges? You could set up your 2-meter/440 MHz radio and work the international space station repeater! Or send a digital stream of information off a meteor trail. Or yak with a pal on the other side of the country through one of the many amateur radio satellites orbiting up there for exclusive use by hams. And for out-of-the-world communications with your new 2-meter ham band Technician Class privileges, you can tune in regularly with the ham radio operator astronauts passing 240 miles overhead in the International Space Station. Yes, the ISS has a complete ham station installed and, when it's overhead, its within line-of-sight range of your handheld (**ariss.org**)!

There's still more you can do with just that entry-level Technician Class license.

If you're also into computers, hams have access to hundreds of wireless frequencies to send high speed information "packets" over the airwaves, absolutely-error-free wireless e-mails, digital slow scan color photos, and binary phase-shift-keying computer-to-computer mode that occupies only a sliver of radio bandwidth called PSK-31.

Live television? As a Technician Class ham radio operator, we can put you on some frequencies where you will join fellow amateur radio television operators to beam high-definition, live-action television all over the state! We call this "ATV."

Imagine talking on your little handheld radio to another ham with their handheld radio *halfway around the world!* Very common these days, thanks to seasoned ham operators who offer free Voice-over-Internet-Protocol gateway station access. Both IRLP and EchoLink® gateway stations are standing by to relay your radio call.

If you're non-technical, we can use a computer to download all the frequencies to your handheld radio memory circuits to turn you into a walkin' talkin' computin' radio operator with equipment not much larger than your favorite cell phone. You don't have to be an engineer to become a ham operator!

Not just a hobby, but a service

There's a serious side to our hobby, when ham radio becomes a public service. When emergencies strike, ham operators are at their shining best. At any major or local disaster, ham operators are often the first to handle emergency calls.

Our ham radio nets stay on the air through hurricanes, during tornadoes, and even in the event of major disasters. Following the 9/11 attacks and the Hurricane Katrina disaster, ham operators worked for more than a month providing additional emergency communications capabilities to the rescue workers. More recently, a team of US hams volunteered for more than 2 months to provide emergency communications support following the devastating hurricanes in Texas, Florida, and Puerto Rico. Our ham radio network of mobile radios, relay stations and remote base equipment continuously keeps emergency responders in contact with many necessary resources.

Many of our ham radio emergency traffic handlers are always at home and always on the air. You wouldn't know that they are visually impaired or perhaps confined to a wheelchair because there is no disability that would keep ham radio operators from working on the amateur radio service airwaves. To learn more, visit **www.handiham.org**.

Join a club – get yourself an "Elmer"

When you pass your upcoming Technician Class ham exam, your local ham radio clubs may send you a letter with a warm welcome inviting you to join them at an upcoming club meeting. You should go!

Club members – who are now your fellow amateur operators – can help you select radio equipment, program your new radio, and even come to your house to help set-up that home or vehicle radio installation. These willing helpers who are eager to "show you the ropes" are known as "Elmers," and your Elmer can teach you the practical, on-the-air aspects of amateur radio. Ham radio is one big fraternity!

What are some other ways to learn about your new hobby? ***CQ – Amateur Radio*** contains excellent articles that will help you learn about all the aspects of the hobby. In addition to the print edition of ***CQ***, they offer a digital edition that contains bonus content, including many articles specifically about Technician Class VHF/UHF operations. Visit **www.cq-amateur-radio.com**. There is a coupon in the back of this book for a free, 3-issue mini-subscription to ***CQ***.

The largest ham radio magazine is ***QST – Amateur Radio***. It is published by the American Radio Relay League (ARRL), which is the national association for amateur radio in the United States (visit **www.arrl.org**). There's a membership application form to join the ARRL in the back of this book, too, which will also entitle you to a free ARRL publication.

And before we let you go on to the next Chapter of our book – which details all of the Technician Class frequency privileges you'll earn when you pass your exam – keep in mind that there are more frequencies and bands that you will earn as you upgrade your ham license. As a General and Extra class operator, you'll gain many more long-range bands to keep you in touch around the world. You could even become a volunteer examiner with the General and Extra class licenses, and then you could give the same exam you are about to pass.

So let's get started now with Technician Class study. In the next Chapter, we'll take an in-depth look at the frequency privileges you'll earn with your Technician Class license. We can't wait to hear you on the air with your new call sign. Welcome to our Amateur Radio service!

2

Technician Class Privileges

There is plenty of excitement out there on the amateur VHF and UHF bands for the ham with a Technician Class license. And thanks to the FCC's ruling that eliminated the Morse code test requirement for all classes of ham radio license, the Technician Class operator also gets a taste of HF excitement on portions of the 80-, 40-, 15-, and 10-meter bands. Some of these band segments are restricted to Morse code only operation, so you'll still want to learn the code, even though it isn't required by the Federal Communications Commission.

SPECTRUM, WAVELENGTH, AND FREQUENCY

Before we look at the actual Technician Class privileges you'll earn with your new FCC Amateur Radio license, let's take a minute to understand the fundamentals of what is meant by the radio terms *spectrum, wavelength,* and *frequency.*

Figure 2-1 on the next page shows the entire electromagnetic energy spectrum, and highlights where the radio frequencies fall within the total spectrum. The low end of the spectrum starts with audio and VLF (Very Low Frequency) frequencies. At the top end of the overall spectrum – above the radio frequencies – are light, X-Rays, and Gamma Rays.

The region from 20,000 hertz to 30 gigahertz is where radio waves are found. Within that region, the radio spectrum is divided up for various uses. The commercial radio AM band is found from 550 kHz (kilohertz) to 1650 kHz. FM radio stations operate between 88 MHz (megahertz) to 108 MHz. One hertz is equal to one cycle per second. That means that an AM signal at 720 kHz on your radio dial is oscillating at 720,000 cycles per second, and an FM signal at 91.5 MHz on your dial is oscillating at 91,500,000 cycles per second!

Figure 2-2 adds some detail to this explanation. The table at the top shows where some of the frequencies lie, and where Amateur Radio operators have privileges. The illustrations show how wavelength and frequency are related. The easiest thing to remember is LOWER LONGER, HIGHER SHORTER. Lower frequency radio waves travel longer distances in one cycle (wavelength), and higher frequency radio waves travel shorter distances in one cycle.

When we say that we are going to operate on the 6-meter band, that means the wavelength of the frequency we will be using is about 6-meters long – or that one cycle of the frequency travels about 236 inches (or 19 feet) in one cycle. On the 70 centimeter band, one wavelength is about 27.5 inches long in one cycle.

In general, antennas need to be equal to wavelength (or ½ or ¼ wavelength) in order to efficiently send and receive radio signals. And, finally, radio waves travel at approximately the speed of light, or 300,000,000 meters per second. So, there's a lot of stuff happening in a big hurry out there on the radio waves!

Now, let's take a look at the Technician Class frequency privileges, and how they are used for various purposes.

Figure 2-1. The electromagnetic spectrum detailing the radio frequency spectrum

Soruce: FCC

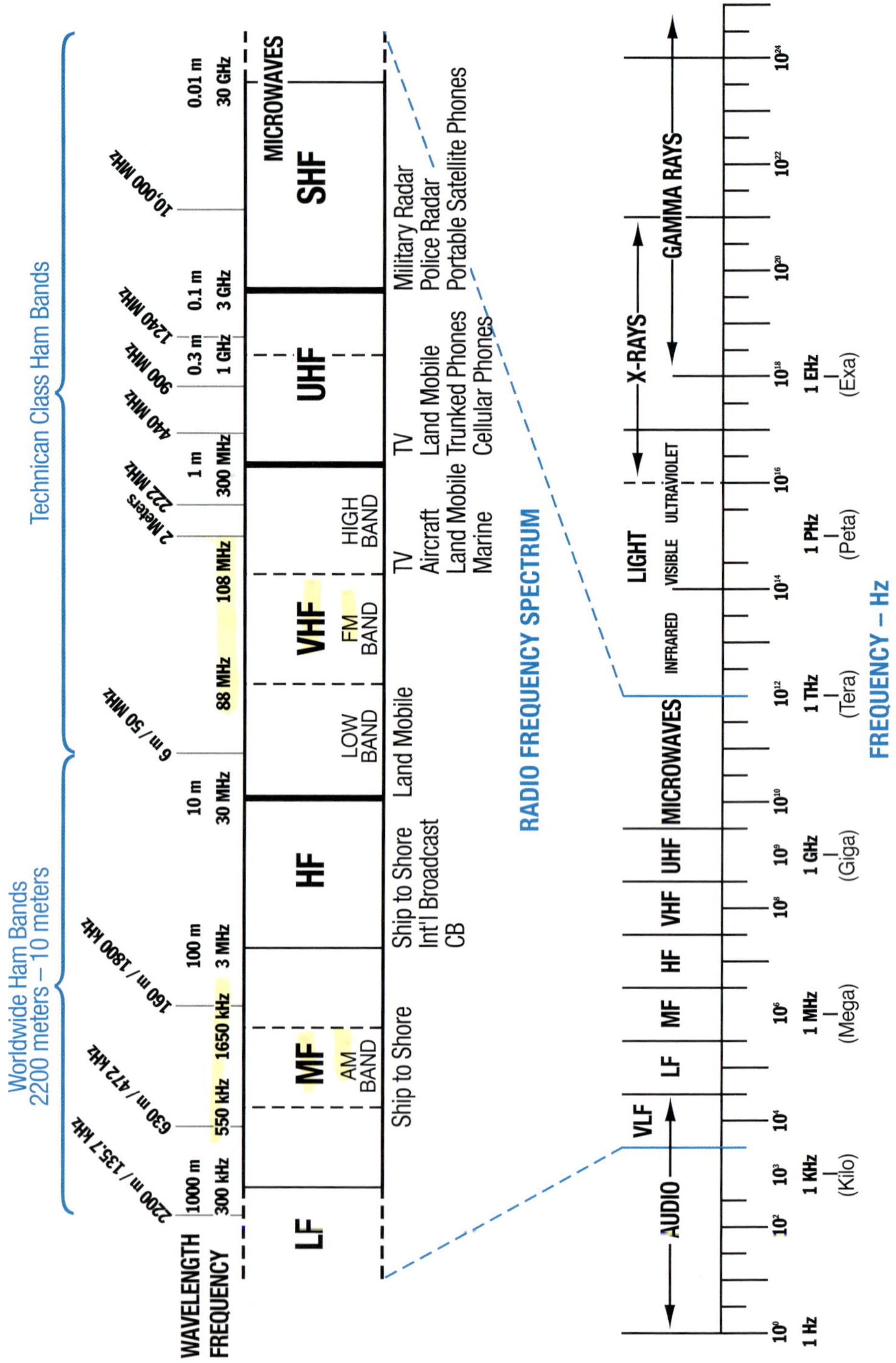

Figure 2-2. Radio Bands, Frequency and Wavelength

Category	Abbrev.	Frequency	Amateur Band Wavelength
Audio	AF	20 Hz to 20 kHz	None
Very Low Frequency	VLF	3 to 30 kHz	None
Low Frequency	LF	30 to 300 kHz	2200 meters
Medium Frequency	MF	300 to 3000 kHz	630, 160 meters
High Frequency	HF	3 to 30 MHz	80, 60, 40, 30, 20, 17, 15, 12, 10 meters
Very High Frequency	VHF	30 to 300 MHz	6, 2, 1.25 meters
Ultrahigh Frequency	UHF	300 to 3000 MHz	70, 33, 23, 13 centimeters
Superhigh Frequency	SHF	3 to 30 GHz	9, 5, 3, 1.2 centimeters
Extremely High Frequency	EHF	Above 30 GHz	6, 4, 2.5, 2, 1 millimeter

Frequency Spectrum

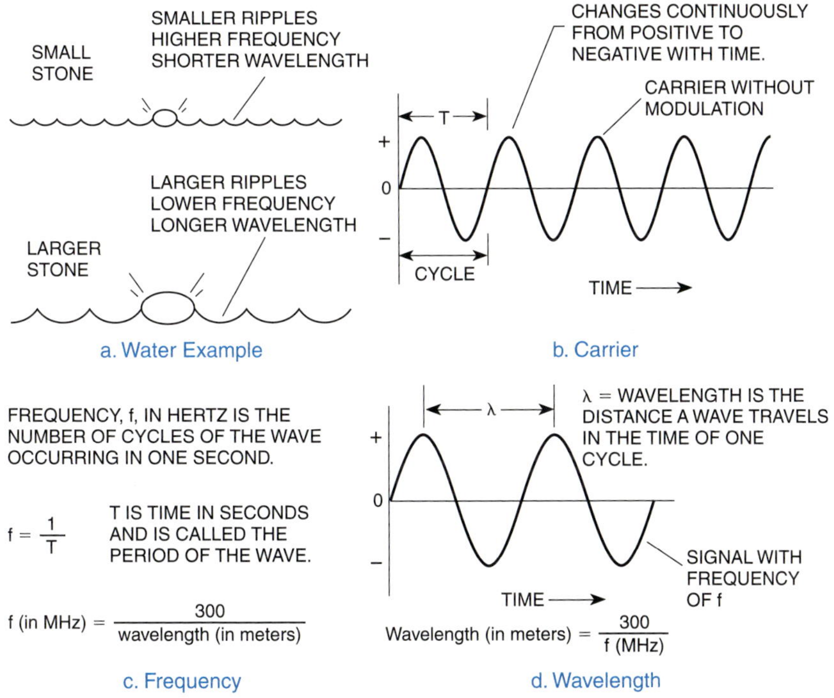

Carrier, Frequency, Cycle and Wavelength

TECHNICIAN CLASS PRIVILEGES

The Technician Class is the entry-level license where you get started in the amateur service. It's now easier than ever to enter the amateur radio service as a Technician Class operator.

As a Technician Class operator, you will have plenty of excitement on radio frequencies above 30 MHz. You will have full operating privileges on all of the exciting VHF, UHF, SHF, and microwave bands shown in *Table 2-1.* You can work skywaves on 6 meters to communicate all over the country; on the 2-meter band, you'll operate through repeaters and orbiting satellites, and on Voice over Internet links. On the 222 MHz band, you may operate through linked repeaters; on the

440 MHz band, you might try amateur television, satellite, and remote base operation; and on 1270 MHz there are more frequencies for amateur television, satellites, repeater linking, and digital messaging and Voice over Internet linking. And then there are the microwave bands where dish, loop, and Yagi antennas will beam out signals a lot farther than you might think.

In addition, you also may operate single sideband voice on a portion of the 10 meter band, where skywaves bounce halfway around the world. You also receive Morse code privileges on portions of the 10-, 15-, 40-, and 80-meter bands. While it's not a requirement to learn the code, or pass a code test, you may enjoy Morse code worldwide operation with your new Technician Class privileges on these 4 worldwide bands.

Table 2-1. Technician Class Operating Privileges

Wavelength Band	Frequency	Emissions	Comments
HF Skywave Operating Privileges			
80 Meters	3525–3600 kHz	Code only	Limited to Morse code (200 watt PEP output limitation)
40 Meters	*7025–7125 kHz	Code only	Limited to Morse code (200 watt PEP output limitation)
15 Meters	21,025–21,200 kHz	Code only	Limited to Morse code (200 watt PEP output limitation)
10 Meters	28,000–28,300 kHz	Code data	Morse code (200 watt PEP output limitation)
	28,300–28,500 kHz	Phone	Sideband voice (200 watt PEP output limitation)
Technician Class Unrestricted VHF/UHF Frequency Privileges			
6 Meters	50.0–54.0 MHz	All modes	Morse code, sideband voice, radio control, FM repeater, digital computer, remote bases, and autopatches (1500 wattsPEP output)
2 Meters	144–148 MHz	All modes	All types of operation including satellite and owning repeater and remote bases. (1500 watt PEP output)
1¼ Meters	219–220 MHz	Data	Point-to-Point digital message forwarding
	222–225 MHz	All modes	All band privileges. (1500 watt PEP output)
70 cm	420–450 MHz	All modes	All band privileges, including amateur television, packet, RTTY, FAX, and FM voice repeaters. (1500 watt PEP output)
33 cm	902–928 MHz	All modes	All band privileges. Plenty of room! (1500 watt PEP output)
23 cm	1240–1300 MHz	All modes	All band privileges. (1500 watt PEP output)
13 cm	2300-2310 MHz 2390-2450 MHz	All modes	Ham T.V. Links; Satellites

* U.S. licensed operators in other than our hemisphere (ITU Region 2) are authorized 7050-7075 kHz due to shortwave broadcast interference.

We will begin by explaining all of your *unrestricted* UHF/VHF privileges, beginning with the 6-meter band.

6-METER WAVELENGTH BAND, 50.0-54.0 MHz

The Technician Class operator enjoys all amateur service privileges and maximum output power of 1500 watts on this worldwide band. Work your computer to send and receive digital modes JT65 and FT8, popular on 6 meters at low power levels. *Table 2-2* shows the 6-meter wavelength band plan.

On 6 meters, the Technician Class operator can get a real taste of long-range skywave skip communications. During the summer months, and during selected days and weeks out of the year, 50-54 MHz, 6-meter signals are refracted by the ionosphere, giving you incredible long-range communication excitement. It's almost a daily phenomenon during the summer months for 6 meters to skip all over the country. This is the big band for the Technician Class operator because of this type of ionospheric, long-range, skip excitement. There are even repeaters on 6 meters. So make 6 meters "a must" at your future operating station.

Table 2-2. 6-Meter Wavelength Band Plan, 50.0-54.0 MHz

MHz	Use
50.000–50.100	CW beacons
50.060–50.080	CW Beacon FM
50.100–50.300	SSB, CW
50.100–50.125	DX window & SSB DX calling
50.125	Domestic SSB calling frequency and QSO each side
50.313 & 50.323	Digital FT8
50.600–50.800	Non-voice
50.620	Digital/Packet calling frequency
50.800–50.980	Radio control
	20 kHz channels
51.000–51.100	Pacific DX window
51.120–51.480	Repeater inputs (19)
51.120–51.180	Digital repeater inputs
51.620–51.980	Repeater outputs (19)
51.620–51.680	Digital repeater outputs
52.000–52.480	Repeater inputs (23)
52.020, 52.040	FM simplex
52.500–52.980	Repeater outputs (23)
52.525, 52.540	FM simplex
53.000–54.480	Repeater inputs (19)
53.000, 53.020	FM simplex
53.1/53.2/53.3/53.4	Radio control
53.500–53.980	Repeater outputs (19)
53.5/53.6/53.7/53.8	Radio control
53.520	Simplex
53.900	Simplex

2-METER WAVELENGTH BAND, 144-148 MHz

The 2-meter band is the world's most popular spot for staying in touch through repeaters. Here is where most all of those handheld transceivers operate, and the Technician Class operator receives unlimited 2-meter privileges! *Table 2-3* gives the 2-meter wavelength band plan adopted by the ARRL VHF/UHF advisory committee.

The United States, and many parts of the world, are blanketed with clear, 2-meter repeater coverage. They say there is nowhere in the United States you can't reach at least one or two repeaters with a little handheld transceiver. 2-meters has you covered! Here are examples:

Handie-talkie channels	Tropo-DX-ducting	Remote base
Transmitter hunts	Internet radio links	Simplex operation
Space station	Contests	Rag-chewing
Moon bounce (EME)	Traffic handling	Emergency nets
Meteor bursts	Satellite downlink	Sporadic-E DX
Packet radio	Satellite uplink	Aurora

The Technician Class license allows 1500 watts maximum power output for specialized 2-meter communications, and also permits you to own and control a 2-meter repeater.

Table 2-3. 2-Meter Wavelength Band Plan, 144-148 MHz

MHz	Use
144.00–144.05	EME (Earth-Moon-Earth) (CW)
144.05–144.10	General CW and weak signals
144.10–144.20	EME and weak-signal SSB plus digital MSK-144 at 144.150 MHz
144.20	National SSB calling frequency
144.20–144.275	General SSB operation, upper sideband
144.275–144.30	Propagation beacons
144.50–144.60	Linear translator inputs
144.60–144.90	FM repeater inputs
144.90–145.10	Weak signal and FM simplex plusD-Star digital
145.10–145.20	Linear translator outputs plus packet
145.20–145.50	FM repeater outputs
145.50–145.80	Miscellaneous and experimental modes
145.80–146.00	OSCAR subband—satellite and space station use
146.01–146.37	Repeater inputs
146.40–146.58	Simplex
146.52	National simplex calling frequency
146.61–146.97	Repeater outputs
147.00–147.39	Repeater outputs
147.42–147.57	Simplex and D-Star systems
147.60–147.99	Repeater inputs

1¼-METER WAVELENGTH BAND, 219-220 MHz & 222-225 MHz

On the 222 MHz band, the frequencies 219 to 220 MHz may be used by point-to-point digital message forwarding stations operated by Technician Class licensees or higher. These stations must register with the ARRL 30 days prior to activation, and must not interfere with primary marine users near the Mississippi River, or any other primary user of this band. The 100 kHz channels for point-to-point fixed digital message forwarding stations, 50 watts PEP limit, are shown in *Table 2-4.*

Table 2-4. 1¼-Meter Wavelength Band Plan, 219-220 MHz

MHz	Use			
219–220	Point-to-point fixed digital message forwarding systems. Must be coordinated through ARRL. 100 kHz Channels. 50W PEP limit.			
	Channel	**Freq. (MHz)**	**Channel**	**Freq. (MHz)**
	A	219.050	F	219.550
	B	219.150	G	219.650
	C	219.250	H	219.750
	D	219.350	I	219.850
	E	219.450	J	219.950

Table 2-5 shows the usage allocation for the 222-225 MHz portion of the band. The Technician Class license permits you to use the entire band at 1500 watts maximum output power. If you need some relief from the activity on 2 meters, the 222-225 MHz band is similar in propagation and use and is exclusively assigned to our Amateur Radio service. 222 MHz equipment is readily available in this country, either as handheld or mobile, with single-, dual- or tri-band capabilities. 222 MHz is a USA band only, with lots of elbow room!

Table 2-5. 1¼-Meter Wavelength Band Plan, 222-225 MHz

MHz	Use
222.00–222.15	Weak-signal modes (FM only)
222.00–222.05	EME (Earth-Moon-Earth)
222.05–222.06	Propagation beacons
222.10	SSB and CW calling frequency
222.10–222.15	Weak signal CW and SSB
222.15–222.25	Local coordinator's option: Weak signal, ACSB, repeater inputs, control points
222.25–223.38	FM repeater inputs only
223.40–223.52	FM simplex
223.50	Simplex calling frequency
223.52–223.64	Digital, packet
223.64–223.70	Links, control
223.71–223.85	Local coordinator's option: FM simplex, packet, repeater outputs
223.85–224.98	Repeater outputs only

70-CM WAVELENGTH BAND, 420-450 MHz

As you gain more experience on the VHF and UHF bands, you will soon be invited to the upper echelon of specialty clubs and organizations. The 440-MHz band is where the experts hang out. *Table 2-6* presents the ARRL 70-cm (centimeter) wavelength band plan. Amateur television (ATV) is very popular, so there's no telling who you may see as well as hear. This band also has the frequencies for controlling repeater stations and base stations on other bands, plus satellite activity. With a Technician Class license, you may even be able to operate on General Class worldwide frequencies if a General Class or higher control operator is on duty at the base control point. You would be able to talk on your 440-MHz handheld transceiver and end up in the DX portion of the 20-meter band. As long as the control operator is on duty at the control point, your operation on General Class frequencies is completely legal!

The 440 MHz band is also full of packet communications, RTTY, and all those fascinating FM voice repeaters. If you are heavy into electronics, you'll hear fascinating topics discussed and digitized on the 440-MHz band. A Technician Class operator has full power privileges as well as unrestricted emission privileges on the 440 MHz band. Visit: **www.winsystem.org**

Table 2-6. 70-cm Wavelength Band Plan, 420-450 MHz

MHz	Use
420.00–426.00	ATV repeater or simplex with 421.25-MHz video carrier control links and experimental
426.00–432.00	ATV simplex with 427.250-MHz video carrier frequency
432.00–432.07	EME (Earth-Moon-Earth)
432.07–432.08	Propagation beacons (old band plan)
432.08–432.10	Weak-signal CW
432.10	70-cm calling frequency
432.10–433.00	Mixed-mode and weak-signal work, plus FT8 (432.500)
432.30–432.40	New beacon band
433.00–435.00	Auxiliary/repeater links
435.00–438.00	Satellite space station contacts only
438.00–444.00	ATV repeater input with 439.250-MHz video carrier frequency and repeater links
442.00–445.00	Repeater inputs and outputs (local option)
445.00–447.00	Shared by auxiliary and control links, repeaters and simplex (local option); (446.0-MHz national simplex frequency)
447.00–450.00	Repeater inputs and outputs

33-CM WAVELENGTH BAND, 902-928 MHz

Radio equipment manufacturers are just beginning to market equipment for this band. Many hams are already on the air using home-brew equipment for a variety of activities. If you are looking for a band with the ultimate in elbow room, this is it! *Table 2-7* shows the 33-cm wavelength band plan adopted by the ARRL.

Table 2-7. 33-cm Wavelength Band Plan, 902-928 MHz

MHz	Use
902.0–903.0	Weak signal (902.1 calling frequency)
903.0–906.0	Digital Communications (903.1 alternate calling frequency)
906.0–909.0	FM repeater inputs
909.0–915.0	ATV
915.0–918.0	Digital Communications
918.0–921.0	FM repeater outputs
921.0–927.0	ATV
927.0–928.0	FM simplex and links

23-CM WAVELENGTH BAND, 1240-1300 MHz

There is plenty of equipment for this band. Technician Class operators may run any legal amount of power—with 20 watts about the usual safe limit. The frequencies are in the microwave region, and this band is excellent to use with local repeaters in major cities. Like the 440-MHz and the 2-meter bands, this band is sliced into many specialized operating areas. You can work orbiting satellites, operate amateur television, or own your own repeater with your Technician Class license. *Table 2-8* presents the 23-cm wavelength band plan adopted by the ARRL.

Table 2-8. 23-cm Wavelength Band Plan, 1240-1300 MHz

MHz	Use
1240–1246	ATV #1
1246–1248	Narrow-bandwidth FM point-to-point links and digital, duplexed with 1258-1260 MHz
1248–1252	Digital communications
1252–1258	ATV #2
1258–1260	Narrow-bandwidth FM point-to-point links and digital, duplexed with 1246-1252 MHz
1260–1270	Satellite uplinks, reference WARC '79
1260–1270	Wide-bandwidth experimental, simplex ATV
1270–1276	Repeater inputs, FM and linear, paired with 1282-1288 MHz, 239 pairs every 25 kHz, e.g., 1270.025, 1270.050, 1270.075, etc. 1271.0-1283.0 MHz uncoordinated test pair
1276–1282	ATV #3
1282–1288	Repeater outputs, paired with 1270-1276 MHz
1288–1294	Wide-bandwidth experimental, simplex ATV
1294–1295	Narrow-bandwidth FM simplex services, 25-kHz channels
1294.5	National FM simplex calling frequency
1295–1297	Narrow bandwidth weak-signal communications (no FM)
1295.0–1295.8	SSTV, FAX, ACSB, experimental
1295.8–1296.0	Reserved for EME, CW expansion
1296.0–1296.05	EME exclusive
1296.07–1296.08	CW beacons
1296.1	CW, SSB calling frequency
1296.4–1296.6	Crossband linear translator input
1296.6–1296.8	Crossband linear translator output
1296.8–1297.0	Experimental beacons (exclusive)
1297–1300	Digital communications

10-GHz (10,000 MHz!) BANDS AND MORE

There are several manufacturers of ham microwave transceivers and converters for this band, so activity is excellent. A transverter tied into your existing rig does the job! Using horn and dish antennas, 10 GHz is frequently used by hams to establish voice communications for controlling repeaters over paths from 20 miles to 100 miles. Output power levels are usually less than one-eighth of a watt! It's really fascinating to see how directional the microwave signals are. If you live on a mountain-top, 10 GHz is for you.

All modes and licensees except Novices are authorized on the bands shown in *Table 2-9.* There is much Amateur Radio experimentation on these bands.

Table 2-9. Gigahertz Bands

2.30–2.31 GHz	10.0–10.50 GHz*	119.98–120.02 GHz
2.39–2.45 GHz	24.0–24.25 GHz	142.0–149.0 GHz
3.30–3.45 GHz	47.0–47.20 GHz	241.0–250.0 GHz
5.65–5.925 GHz	75.50–81.0 GHz	All above 300 GHz

*Pulse not permitted

HIGH-FREQUENCY (HF) TECHNICIAN CLASS PRIVILEGES

So, you're about to take your Technician Class license exam. Congratulations! If you pay attention to our excellent advice and wisdom all through this book, you'll have a pretty good idea about what to do *after* you get your license. As a Technician Class licensee, you will have some amazing operating privileges and possibilities at your fingertips, or typically, on your hip!

Kids make great ham radio operators.

We'd like to talk to you for a few moments about some privileges you have as a Technician Class licensee that are often ignored, your HF privileges. These are your frequencies between 3 and 30 MHz that were "grandfathered" to you from the former Novice class license, which not that long ago was the entry level Amateur Radio license.

As a Technician Class licensee, you have Morse code (CW) privileges on 80, 40, and 15 meters, plus CW *and* voice privileges on 10 meters. This is in addition to all the VHF/UHF privileges that we just explained on the preceding pages. These HF frequencies give you *international* communications capabilities, far beyond the normal line-of-sight propagation afforded by typical VHF and UHF operation. Although great distances *can* be achieved on VHF/UHF (and Gordo has written a great deal about this), this is not the *normal* mode of operation for most hams. On HF, long-distance communications *is* the normal thing to do, and it opens up a whole new world, quite literally, of amateur radio excitement.

Now, here's the one fly in the ointment. To use 80, 40, or 15 meters (and the lower part of 10 meters), you have to know Morse code! Well, that's not entirely true any more. You have to *use* Morse code (more accurately known as the International Radiotelegraph Code) on those bands. There are many software (and hardware) tools available that allow you to transmit and receive Morse code with a computer without having to know a single character of Morse code. But why do that? Learning Morse code is fun and easy! Just because you aren't required to learn Morse code to get a ham license doesn't mean you won't *want* to learn it. Gordo has spent many years teaching Morse code, and his code teaching materials have long been a mainstay of his excellent teaching materials. Many hams find Morse code one of the most enjoyable aspects of ham radio. Just tuning around the HF ham bands you will find an incredible amount of CW activity. Countless hams still use CW; not because they have to, but because they *want* to.

For the new ham, the real advantage of CW is that the equipment is amazingly inexpensive. You can build a complete CW transceiver, such as a Rockmite, which will give you low power (QRP) capabilities on a single band, such as 40 meters, for as little as $40. Under the right conditions, you can work the world with a Rockmite. Other similar kits are available as well, or you might try "rolling your own" with the countless "homebrew" projects available in many ham radio publications.

Straight Key

You might also want to check out SKCC (Straight Key Century Club) activities. SKCC is a relatively new fraternity within ham radio, composed of hams who have rediscovered the joys of using a *straight key* to transmit Morse Code. Most of the activity on the SKCC net frequencies is slow, and a great way to break into CW with a minimum of terror. Speaking of terror, most new hams experience a great deal of "mike fright" when first getting on the air. One of the unheralded advantages of CW is its high degree of anonymity! Until folks get to know you, you're just a string of dots and dashes somewhere out in the ether!

If the prospect of CW *still* doesn't seem your cup of tea, you can plunge into HF on 10 meters where you do have phone (voice) privileges. 10 Meters is a fascinating band, capable of extremely long distances with very low power, and full of surprises. Several 10 meter-only SSB (Single Sideband) transceivers are available at very reasonable prices. And, in all likelihood, you can borrow a spare "boat anchor" which has 10 meter capabilities from a local *Elmer,* until you decide that you need one of your own. There is even an entire society within ham radio dedicated entirely to 10 meter operation, the 10-10 club, sponsored by Ten-Ten International. There are some 70,000 10-10 members, all dedicated to promoting activity on 10 meters. To become a member, all you have to do is confirm 10 meter contacts with 10 other 10-10 members, after which you will receive a lifetime "10-10 number." Eric's is 34364 and is almost as much a part of his identity as his callsign! Gordon's is 17366.

One of the surprises on 10 meters is *Sporadic E* propagation, which can occur nearly any time of the year and at *any time of day or late afternoon*. Sporadic E can afford startlingly loud signals from nearly anywhere in the world. This is something that rarely occurs even on the "better" HF bands on which you have privileges. Even when unusual 10-meter propagation isn't in effect, 10 meters is a *very* reliable band for local and regional operation, such as emergency communications, especially since it does not rely on an often fragile and vulnerable repeater system, as do VHF and UHF. Ten meter *mobile* operation has a long and fun history in Amateur Radio. Try digital modes, too. "WSJT."

Don't ignore your HF privileges! They are far more than a mere afterthought. Once you check them out, you'll probably find that they are your most *valuable* privileges!

80-METER WAVELENGTH BAND, 3500-4000 KHz

Your privileges on the 75/80-meter band are for CW only from 3525 kHz to 3600 kHz.

40-METER WAVELENGTH BAND, 7000-7300 KHz

You will have Morse-code-only privileges on this band from 7025 kHz to 7125 kHz. This is a popular nighttime and early morning band because signals in code can reach up to 5,000 miles away!

15-METER WAVELENGTH BAND, 21,000-21,450 KHz

Technician Class operators may operate Morse code from 21,025 kHz to 21,200 kHz in this portion of the worldwide band. You can expect daytime range in excess of 10,0000 miles using CW.

10-METER WAVELENGTH BAND, 28,000-29,700 KHz

You may operate code and digital computer communications from 28,000 kHz to 28,300 kHz, and monitor 28.2 MHz to 28.3 MHz for low-power propagation beacons. Now here's the good news – you may operate single-sideband voice between 28.3 MHz to 28.5 MHz, and literally work the world during daylight hour band openings! 10 meters is a great band to introduce you to the excitement of long-range, worldwide voice operation on the HF bands. If you have a CB antenna, it will work very well for Technician Class privileges on 10 meters!

If you're joining the ham radio hobby from CB radio operation on 11 meters, you know all the excitement of skywave propagation. Your Technician Class 10 meter privileges allow you to own and operate a big worldwide radio, and enjoy ionospheric skip at its best! As we climb to the peak of solar cycle 25, conditions on 10 meters will constantly surprise the dayllights out of you!

DOWNLOAD THE ON THE AIR AUDIO CLASS NOW!

Our "On the Air" audio class is a one-hour long introduction to all the excitement that awaits you when you get on the air with your new Technician Class privileges. Download it now from **w5yi.org/page.php?id=367** and listen to hear:

Track 1 introduces you to the worldwide HF bands that your Technician license authorizes. Talk thousands of miles using skip communications on the 10 meter band.

Track 2 introduces you to skywave contacts you can achieve on the 6 meter band.

Track 3 lets you listen in on everything the 2 meter band has to offer the new Technician Class operator. You can even talk to the International Space Station astronauts on 2 meters!

Track 4 puts you on the 222 MHz band for special event stations and more.

Track 5 tunes you in to some exciting satellite communications, plus ham color TV!

Track 6 takes you to the top bands for Technician Class radio excitement. Imagine communicating on frequencies 10 times higher that your microwave oven!

Gordo recorded this audio course to acquaint you with all the sounds of the airwaves. Everything you hear is something you will be able to do with your new Technician Class license!

LEARNING MORE ABOUT HAM RADIO

How do you get to be a knowledgeable amateur radio operator – one who understands how radios work and the ins-and-outs of our hobby? One of the best ways to learn all about your new avocation is by reading the amateur radio magazines. One of Gordo's favorites is ***CQ – Amateur Radio***. Every month, ***CQ*** is full of a range of articles that will give you operating tips, show you how radio technology works, and looks at new products. ***CQ*** always includes a column, Beginner's Corner, to help you get started in ham radio.

To help you get started with ***CQ***, we've included a special offer in the back of this book. It is a coupon good for a 3-issue, mini-subscription to ***CQ*** – that's **FREE** 3 issues of the magazine to help you get started in ham radio. All you have to do is clip out the order form, fill in your name and address, and drop it in a nearby mailbox. So give ***CQ*** a try. There's a lot of good info in it every month!

AN IMPORTANT WORD ABOUT SHARED FREQUENCIES

In this Chapter, we discussed the Amateur Radio frequency privileges you will receive when you pass your exam and receive your license and call sign from the FCC. It is important, however, that you know that every ham band above 225 MHz is shared on a secondary basis with other services. This means that the primary users get first claim to the frequency!

For example, Government radio location (RADAR) is a primary user of some bands. And a multitude of industrial, scientific and medical services have access to the 902-928 MHz band. Just because the frequency is allocated to the amateur service does not mean that others do not have prior right or an equal right to the spectrum. You must not interfere with other users of these bands.

Amateurs must not cause interference to other stations such as foreign stations operating along the Mexican and Canadian borders, military stations near military bases, and FCC monitoring stations. Also, amateur operators must not cause interference in the National Radio Quiet Zones which are near radio astronomy locations. The astronomy locations are protected by law from Amateur Radio interference. Operation aboard ships and aircraft is restricted. The FCC can curtail the hours of your operation if you cause general interference to the reception of telephone or radio/TV broadcasting by your neighbors.

Every amateur should have a copy of the Amateur Radio Service Part 97 Rules and Regulations. It would be good for you to read Part 97.303 on frequency sharing.

SUMMARY

Enjoy the excitement of being a ham radio operator and communicating worldwide with other amateur operators. Your Technician Class operator license allows you all ham operating privileges on all bands with frequencies greater than 50 MHz. You have operating privileges on the worldwide 6-meter band, on the world's popular 2-meter and 222-MHz repeater bands, on the amateur television, satellite communications, and repeater-linking 440-MHz and 1270-MHz bands, and on the line-of-sight microwave bands at 10 GHz and above.

Remember, you enter the amateur service as a Technician Class operator just by passing the Element 2 written exam. You will receive full privileges on all of the VHF/UHF bands we have just described, along with the HF privileges on 80-, 40-, 15-, and 10-meters.

MOST IMPORTANT, ask your examination team members how to join the local amateur radio club. Fellow ham members will welcome you to our fraternity and give you a hand with all the exciting new things you will experience in ham radio.

Get on the air! The Technician Class license is designed to get you "up and running" in a hurry, so you'll want to take advantage of your new operating privileges as soon as you can get your hands on a rig.... or two.... or three!

All of the questions and answers you are about to study will begin to make sense once you begin operating on the air! A simple dual-band handheld, covering 2 meters and 70 cm, will pull in exciting radio communications that you will certainly enjoy. You could even talk with an astronaut aboard the International Space Station with that little handheld transceiver!

Now, let's take a look at a little ham radio history – read on.

3

A Little Ham History!

Ham radio has changed a lot in the 100+ years since radio's inception. In the past 25 years, we have seen some monumental changes! So, before we get started preparing for the exam, we're going to give you a little history lesson about our hobby, its history, and an overview of how you'll progress through the amateur ranks from your first, entry-level Technician Class license to the top amateur ticket – the Amateur Extra Class license. We know this background knowledge will make you a better ham!

In this chapter you'll learn all of the licensing requirements under the FCC rules that became effective April 15, 2000. And you'll learn about the six classes of license that were in effect *prior* to those rules changes. That way, when you run into a Novice or Advanced class operator on the air, you'll have some understanding of their skill level, experience, and frequency privileges.

WHAT IS THE AMATEUR SERVICE?

There are more than 842,000 licensed amateur radio operators in the U.S. today. The Federal Communications Commission, the Federal agency responsible for licensing amateur operators, defines the amateur radio service this way:

"The amateur service is for qualified persons of all ages who are interested in radio technique solely with a personal aim and without pecuniary interest."

Ham radio is first and foremost a fun hobby! In addition, it is a service. Note the word "qualified" in the FCC's definition – that's the reason you're studying for an exam; so you can prove by passing the exam that you are qualified to get on the air.

Millions of operators around the world exchange ham radio greetings and messages by voice, teleprinting, telegraphy, facsimile, and television worldwide. Japan, alone, has more than three million hams! It is very commonplace for U.S. amateurs to communicate with Russian amateurs, while China is just getting started with its amateur service. Being a ham operator is a great way to promote international good will.

The benefits of ham radio are countless! Ham operators are probably known best for their contributions during times of disaster. In recent years, many recreational sailors in the Caribbean who were attacked by modern-day pirates have had their lives saved by hams directing rescue efforts. Following the 9/11 terrorist attacks on the World Trade Center and the Pentagon, thousands of local hams assisted with emergency communications. In addition, over the years, amateurs have contributed much to electronic technology. They have even designed and built their own orbiting communications satellites.

The ham community knows no geographic, political or social barrier. If you study hard and make the effort, you can be part of our fraternity. Follow the suggestions in our book and you will be well prepared to pass the written exam.

A BRIEF HISTORY OF AMATEUR RADIO LICENSING

Government licensing of radio stations and amateur operators began with The Radio Act of 1912, which mandated the first Federal licensing of all radio stations and assigned amateurs to the short wavelengths of less than 200 meters. These "new" requirements didn't deter them, and within a few years there were thousands of licensed ham operators in the United States.

Since electromagnetic signals do not respect national boundaries, radio operators are international in scope. National governments enact and enforce radio laws within a framework of international agreements which are overseen by the International Telecommunications Union. The ITU is a worldwide United Nations agency headquartered in Geneva, Switzerland. The ITU divides the radio spectrum into a number of frequency bands, with each band reserved for a particular use. Amateur radio is fortunate to have many bands allocated to it all across the radio spectrum.

In the U.S., the FCC is the government agency responsible for regulation of wire and radio communications. The FCC further allocates frequency bands to the various services – including the Amateur Service – in accordance with the ITU plan and regulates stations and operators.

In the early years of amateur radio licensing in the U.S., the classes of licenses were designated by the letters "A," "B," and "C." The highest license class with the most privileges was "A." In 1951, the FCC dropped the letter designations and gave the license classes names Novice, Technician, General and Extra. In 1967, the Advanced class was added. The Novice and Technician exams required 5-wpm code speed. The General exam required 13-wpm code speed, and Extra required 20-wpm. Each of the five written exams were progressively more comprehensive and formed what came to be known as the *Incentive Licensing System.*

In 1979, the international Amateur Service regulations were changed to permit all countries to waive the manual Morse code proficiency requirement for "...stations making use exclusively of frequencies above 30 MHz." This set the stage for the creation of the Technician "no-code" license in 1991, when the 5-wpm Morse code requirement for the Technician Class was eliminated.

By this time, there was a total of six Amateur Service license classes – Novice, Technician, Technician-Plus, General, Advanced, and Extra – along with five written exams and three Morse code tests used to qualify hams for their various licenses.

The Amateur Service Is Restructured

Following an extensive review begun in 1998, the FCC implemented a complete restructuring of the U.S. amateur service that became effective April 15, 2000. Today, applicants can only be examined for three amateur license classes:

- Technician Class – the VHF/UHF entry level license;
- General Class – the HF entry level license, and
- Amateur Extra Class – a technically-oriented senior license.

Most recently, the FCC updated its amateur radio rules by eliminating the Morse code test requirement for all classes of license. The new rule went into effect in February, 2007. Now you can earn any class of license – Technician, General, and Extra Class – simply by passing written exams.

Individuals with licenses issued before April 15, 2000, were "grandfathered" under the new rules. This means that Novice and Advanced class amateurs are able to modify and renew their licenses indefinitely. Technician-Plus amateur licenses were renewed as Technician Class. The FCC elected not to change the operating privileges of any class, so you may hear some of these "grandfathered" hams when you get on the air.

Self-Testing In The Amateur Service

Prior to 1984, all amateur radio exams were administered by FCC personnel at FCC Field Offices around the country. In 1984, the VEC (Volunteer Examiner Coordinator) System was formed after Congress passed laws that allowed the FCC to accept the services of Volunteer Examiners (or VEs) to prepare and administer amateur service license examinations. The testing activity of VEs is managed by Volunteer Examiner Coordinators (or VECs). A VEC acts as the administrative liaison between the VEs who administer the various ham examinations and the FCC, which grants the license.

A team of three VEs, who must be approved by a VEC, is required to conduct amateur radio examinations.

In 1986, the FCC turned over responsibility for maintenance of the exam questions to the National Conference of VECs, which appoints a Question Pool Committee (QPC) to develop and revise the various question pools according to a schedule.

That completes your history lesson. Now let's turn our attention to the privileges you'll earn as a new Technician Class operator.

LICENSE PRIVILEGES

An amateur operator license conveys many privileges. As the control operator of an amateur radio station, you will be responsible for the quality of the station's transmissions. Most radio equipment must be authorized by the FCC before it can be widely used by the public but, for the most part, this is not true for amateur equipment!

Licensed amateur radio operators may design, construct, modify and repair their own equipment. But you must have a license to do this, and even though it is easier than ever, there are certain things you must know before you can obtain your license from the FCC. Everything you need to know is covered in this book.

OPERATOR LICENSE REQUIREMENTS

To qualify for an amateur operator/primary station license, a person must pass an examination according to FCC guidelines. The degree of skill and knowledge that the candidate demonstrates to the examiners determines the class of operator license for which the person is qualified.

Anyone is eligible to become a U.S. licensed amateur operator, including foreign nationals, if they are not a representative of a foreign government. There is no age limitation – if you can pass the examinations, you can become a ham!

OPERATOR LICENSE CLASSES AND EXAM REQUIREMENTS

Today, there are three amateur operator licenses issued by the FCC – Technician, General, and Extra. Each license requires progressively higher levels of learning and proficiency and each gives you additional operating privileges. This is known as *incentive licensing* – a method of strengthening the amateur service by offering more radio spectrum privileges in exchange for more operating and electronic knowledge.

There is no waiting time required to upgrade from one amateur license class to another; nor any required waiting time to retake a failed exam. You can even take all three examinations at one sitting if you're really prepared! *Table 3-1* details the amateur service license structure and required examinations.

Table 3-1: Current Amateur License Classes and Exam Requirements
(Effective April 15, 2000)

License Class	Exam Element	Type of Examination
Technician Class	2	35-question, multiple-choice written examination. Minimum passing score is 26 questions answered correctly (74%).
General Class	3	35-question, multiple-choice written examination. Minimum passing score is 26 questions answered correctly (74%).
Extra Class	4	50-question, multiple-choice written examination. Minimum passing score is 37 questions answered correctly (74%).

ABOUT THE WRITTEN EXAMS

What is the focus of each of the written examinations, and how does it relate to gaining expanding amateur radio privileges as you move up the ladder toward your Extra Class license? *Table 3-2* summarizes the subjects covered in each written examination element.

Table 3-2. Question Element Subjects

Exam Element	License Class	Subjects
Element 2	Technician	Elementary operating procedures, radio regulations, and a smattering of beginning electronics. Emphasis is on VHF and UHF operating.
Element 3	General	HF (high-frequency) operating privileges, amateur practices, radio regulations, and a little more electronics. Emphasis is on HF bands.
Element 4	Extra	Basically a technical examination. Covers specialized operating procedures, more radio regulations, formulas and heavy math. Also covers the specifics on amateur testing procedures.

No Jumping Allowed

All written examinations for an amateur radio license are additive. You *cannot* skip over a license class or bypass a required examination as you upgrade from Technician to General to Extra. You must pass each one before attempting the next.

TAKING THE ELEMENT 2 EXAM

Here's a summary of what you can expect when you take the Element 2 written examination for you Technician Class license. You can take your exam at an in-person test site, or at a remote on-line test session via a Zoom connection. Detailed information about how to find an exam session, what to expect at the session, what to bring to the session, and more, is included in Chapter 5.

All amateur radio examinations are administered by a team of at least three Volunteer Examiners (VEs) who have been accredited by a Volunteer Examiner Coordinator (VEC). The VEs are licensed hams who volunteer their time to help our hobby grow. Examination sessions are organized under the auspices of an approved VEC. A list of the 14 FCC-authorized VEC organizations is in the Appendix on page 218. Five VEC organizations offer both in-person and remote on-line exam sessions.

The Element 2 written exam is a 35-question, multiple-choice test:

- At an **in-person session**, the VEs will give you a test paper that contains the 35 questions and multiple-choice answers, as well as an answer sheet for you to complete. Once you're finished, double check your work before handing in your exam papers. The VEs will score your exam immediately. You'll know before you leave the exam site whether you've passed.
- At a **remote on-line session**, you will log on via Zoom to the VEs who are hosting the exam. Once you register, your exam will appear on your computer screen. You select your answer choices and the ExamTools software records your answers. When you finish, the computer will score your exam and you'll know very quickly if you passed. The remote on-line option allows you to take your exam on your computer from your home, your office, or anywhere you have access to an internet connection.

Whether you take your exam in-person or on-line, take your time! Make sure you read each question and all the answer choices carefully before selecting an answer. Chances are very good that, if you've studied hard, you'll get that passing grade!

HOW MANY CLASSES OF LICENSES?

Once you've passed the Element 2 exam and go on the air as a new Technician Class operator, you'll be talking to fellow hams throughout the U.S. and around the world. Here's a summary of the new and "grandfathered" licenses that your fellow amateurs may hold.

New License Classes

Following the FCC's restructuring of Amateur Radio that took effect April 15, 2000, there are just three license classes – Technician, General, and Extra.

"Grandfathered" Licensees

Once you get on the air with your new Technician Class privileges, you might have a contact with a Novice operator while talking on 10 meters, or while sending CW on 15, 40, or 80 meters. And you might bump into an Advanced Class operator. These Novice and Advanced Class operators get to keep their FCC license class designation even though those licenses are no longer issued.

GETTING YOUR FIRST CALL SIGN

Once the VE team scores your exam and you've passed, the process of getting your official FCC Amateur Radio License begins – usually that same day.

At the exam site, or on-line, you will complete NCVEC Form 605, which is your application to the FCC for your license. If you pass the exam, the VE team will send the required paperwork on to their VEC. The VEC reviews the paperwork and then files your application with the FCC. This filing is done electronically.

FCC Now Charges Licensing Fee

In April 2022, the FCC began charging a new $35.00 application fee for Amateur Radio licenses. Previously, licenses were granted free of charge. This FCC fee is in addition to the examination fee paid to the VE team.

When the FCC receives your examination information from the VEC and approves the application, you will receive an email with a link to the FCC Fee Payment Portal and payment instructions. You will pay the fee using your credit or debit card.

You have 10 days from the date of the email to pay the fee. Make sure to monitor your e-mail inbox and spam folder so you can pay your application fee promptly.

Alternatively, if you are unable to monitor your email inbox or spam folder, you can go on the FCC website **wireless2.fcc.gov/UlsApp/UlsSearch/searchLicense.jsp** and search for your approved application using your FRN number. When you see your pending application that has been approved by the FCC there, you can follow the instructions and to go to the FCC Fee Payment Portal to pay the $35.00 fee.

Either way, once your fee is paid, the FCC will issue your license grant, usually within a few hours on normal business days.

Next, you will receive a second email from the FCC with a link directly to your official license record. ***The link is good for 30 days.*** You also will be able to view, print, and download an official copy of your license by logging into your FCC ULS account.

Your ULS FRN search on the FCC website will look like this.

The FCC discontinued mailing or providing printed licenses in February 2015. Licensees can log into ULS with their FRN and password at any time to view and manage their license record, including updating their email or mailing address.

Your New Call Sign

Your first call sign is assigned by the FCC computer and you have no choice of letters. However, once you have that call sign, you can apply for a Vanity Call Sign of your choice.

Make sure to read Chapter 5 for more details on how this new system will work.

GET YOUR FRN BEFORE THE EXAM

Another change by the FCC requires all applicants to have an FRN – a Federal Registration Number. The FCC no longer allows you to use your Social Security Number as identification. This change was made to help protect the identity of license applicants.

You obtain an FRN by going to the FCC's website:

apps.fcc.gov/cores/userlogin.do

Again, make sure to read Chapter 5 for more details on how this all works.

RE-GAINING YOUR PRIOR PRIVILEGES

Recent FCC rules changes make it possible for hams whose licenses have expired to re-gain their privileges simply by taking and passing the Element 2 Technician Class exam. This rule applies to those with expired General, Advanced, and Extra Class licenses that are beyond the 2-year grace period and appear as CANCELLED in the ULS database.

Special case: Individuals who held a Technician Class license prior to March 21, 1987, are eligible to upgrade to General Class simply by passing the current Technician Class Element 2 examination. The FCC does not require a person to have been continuously licensed to qualify for this upgrade.

Other previously licensed hams can regain their expired Technician, General, or Amateur Extra Class privileges simply by passing the current Element 2 exam, without taking any additional examinations. If you are taking the Element 2 exam to regain your expired license privileges you must present documentation showing that you were previously licensed to receive Element credit. When you pass the current Element 2 Technician Class exam, if you held:

- Novice Class license, no Element credit is allowed and you must pass the exam to be issued a new Technician Class license.
- General Class license, you will receive Element 3 credit for a new General Class license.
- Advanced Class license, you will receive Element 3 credit for a new General Class license.
- Amateur Extra Class license, you will receive Elements 3 and 4 credit for a new Extra Class license.

You will need to bring to the exam session the following to demonstrate your previous license credit: either the original or a copy of your expired license; or proof of holding an expired license. Acceptable forms of proof of an expired license include the following:

- A printout from the QRZ.com old 1993 database showing your General, Advanced, or Extra license.
- A photocopy of an ***Amateur Radio Call Book*** page showing your General, Advanced, or Extra license.
- Verification letter from a VEC showing General, Advanced, or Extra license that the VEC has verified from the FCC database.

- A photocopy of a pre-1987 ***Amateur Radio Call Book*** page (where no license classes were shown) showing your call sign as follows:
- 1x3 – K or W prefix indication a Technician, General, or Advanced license to receive Element 3 General Class only, or
- 2x3 – WA or WB prefix indicating a Technician, General, or Advanced license to receive Element 3 General Class only.
- Any – Element credit for Extra Class MUST show the license class or the applicant will receive General Class element credit only.

If you need documentation:

- Try searching the QRZ.com 1993 Callbook at **qrz.com/search1993.html**
- The W5YI-VEC can search its library of ***Amateur Radio Call Books*** for proof of your expired call sign. It charges a nominal fee for providing this service and mailing you the acceptable proof of license if they are able to locate your information from the various reliable resources. Call them at 800-669-9594 for more information.
- Contact Amateur Radio call sign historian Pete, NL7XM, at TwelveVDC@aol.com

IT'S EASY!

The primary pre-requisite for passing any amateur radio operator license exam is the will to do it. If you follow the suggestions in our book, your chances of passing the Technician exam are excellent.

Yes, indeed, the year 2000 brought some big changes to ham radio. Everyone comes out a winner! There has never been a better time to join the ranks of ham radio hobbyists. So study hard! We hope to hear you on the air very soon.

There's a world of excitement waiting for you when you pass the Technician Class exam!

Want to find a test site fast?
Visit the W5YI-VEC website at **www.w5yi.org**, or call 800-669-9594.

4

Getting Ready for the Exam

The Technician Class written examination consists of 35 multiple-choice questions taken from the 411 questions that make up the 2022-26 Element 2 pool. Each question on your examination and the multiple-choice answer will be identical to what is contained in this book.

This chapter contains the official, complete 411-question FCC Element 2 Technician Class question pool from which your examination will be taken. Again, the exam will contain 35 of these questions, and you must get 74% of the questions correct – which means you must answer 26 questions correctly in order to pass. This chapter also contains important information about how the exam is constructed using the questions taken from the Element 2 pool.

Your examination will be administered by a team of three or more Volunteer Examiners (VEs) – amateur radio operators who are accredited by a Volunteer Examiner Coordinator (VEC). You will receive a Certificate of Successful Completion of Examination (CSCE) when you pass the examination. This is official proof that you have passed. It will be given to you before you leave the exam center. In just a few days, you'll find your call sign on the FCC website, and as soon as you find it there you are licensed to go on the air!

THE 2022-2026 QUESTION POOL

Development and maintenance of all three of the Amateur Radio Question Pools is the responsibility of the National Conference of Volunteer Examination Coordinators (NCVEC). This important job was assigned to the NCVEC by the Federal Communications Commission in 1986. The NCVEC appoints a Question Pool Committee from its membership to review and update the Technician, General, and Extra class question pools. Each pool is updated once every 4 years.

The exam questions in this 2022-26 Element 2 Technician Class Question Pool were authored by a team of five active amateur radio operators. Each of these active hams helped form this new pool by removing old questions on such topics as obsolete technology; adding new questions on such topics as the newest operating techniques and technology, and current rules; and, most important, selecting questions they feel a brand new ham radio operator needs to know to get on the air successfully!

The NCVEC Question Pool Committee members who spent many months working on this new Technician Class question pool are: Roland Anders, K3RA, Anchorage VEC, Chairman of the QPC; and committee members Michael J. Mastroleo, AJ6NJ, GLAARG VEC; Larry Pollock, NB5X, W5YI VEC; and Maria Somma, AB1FM, ARRL VEC.

WHAT THE EXAMINATION CONTAINS

The examination questions and the multiple-choice answers (one correct answer and three "distracters") for all license class levels are public information. They are widely published and are identical to those in this book. *There are no "secret" questions.* FCC rules prohibit any examiner or examination team from making any changes to any questions, including any numerical values. No numbers, words, letters, or punctuation marks can be altered from the published question pool. The 411 Element 2 questions in this book are the same exact questions that will appear on your 35-question Element 2 written examination. But which 35 out of the 411 total questions?

Table 4-1 shows how the Element 2, 35-question examination will be constructed. The question pool is divided into 10 subelements. Each subelement covers a different subject and is divided into topic areas. For example, on your Element 2 examination, of the 48 total pool questions in Subelement T8 on radio activities, 4 questions will be on your exam. Are you a little rusty on junior-high-school-level division and multiplication? Out of the 57 total pool questions on Ohm's Law and power calculations, only FOUR QUESTIONS that could deal with simple math will appear on your exam!

Table 4-1. FCC Element 2 Technician Class Question Pool

Subelement	Topic	Total Questions	Exam Questions
T1	Commission's Rules	67	6
T2	Operating Procedures	36	3
T3	Radio Wave Propagation	34	3
T4	Amateur Radio Practices	24	2
T5	Electrical Principles	52	4
T6	Electronic & Electrical Components	47	4
T7	Practical Circuits	43	4
T8	Signals & Emissions	48	4
T9	Antennas & Feed Lines	24	2
T0	Safety	36	3
TOTALS		411	35

All Volunteer Examination teams use the same multiple-choice question pool. This uniformity in study material ensures common examinations throughout the country. Most exams are computer-generated, and the computer selects one question from each topic area within each subelement for your upcoming Element 2 exam.

Trust us, every question on your upcoming Element 2 exam will look very familiar to you by the time you finish studying this book

QUESTION CODING

Each and every question in the 411 Element 2, Technician Class question pool is numbered using a **code.** *The coded numbers and letters reveal important facts about each question!*

The numbering code always contains 5 alphanumeric characters to identify each question. Here's how to read the question number so you know exactly how the examination computer will select one question out of each group for your exam. Once you know this information, you can increase your odds of achieving a "max" score on the exam, especially if there is a specific group of questions which seems impossible for you to memorize or understand. When you get to Element 4, Extra Class – a very tough exam – this trick will really come in handy!

Let's pick a typical question out of the pool – T8A03 – and show you how this numbering code works:

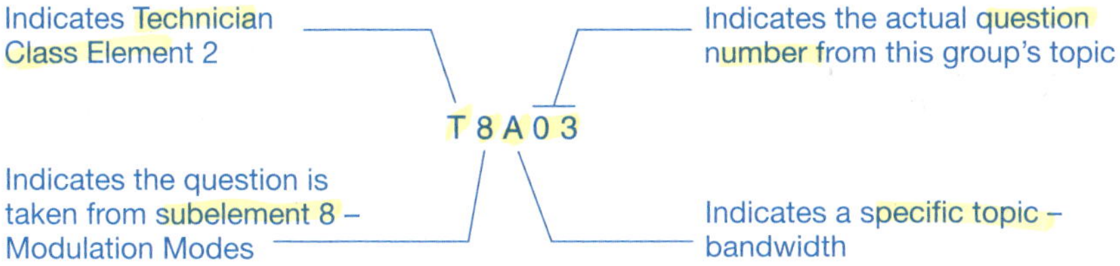

Figure 4-1. Examination Question Coding

- The first character "T" identifies the license class question pool from which the question is taken. "T" is for Technician. "G" would be for General, and "E" would be for Extra.
- The second digit, an "8", identifies the subelement number, 1 through 0. Technician subelement 8 deals with modulation modes, amateur satellite operations, operating activities, non-voice communications.
- The third character, "A", indicates the topic area within the subelement. Topic "A" deals with FCC modulation modes, bandwidth, and emission types.
- The fourth and fifth digits indicate the actual question number within the subelement topic's group. The "03" indicates this is the third question about the use of single sideband modulation. There are 12 individual questions in topic area T8A, *but only one out of this topic group will appear on the test.*

Here's the Secret Study Hint

Only one exam question will be taken from any single group! A computer-generated test is set up to take one question from one single topic group. It cannot skip any one group, nor can it take any more than one question from that group.

Your upcoming Element 2 Technician Class written exam is relatively easy with no bone-crusher math formulas. The same thing is true for the Element 3 exam, General Class. But when you get to the *Extra Class* book, there may be one or two groups that have formulas so complicated that you may want to wait until the very end to digest them. And if you decide to skip them completely, guess what – how many questions out of any one group? That's right, only one per group. This means you are not going to get hammered on any upcoming test with a whole bunch of questions dealing with a specific group topic. Great secret, huh?

Study Time

The Technician Class question pool in this book is valid from July 1, 2022, through June 30, 2026. This pool contains 411 questions written to a middle-school reading level to make sure applicants with minor reading disabilities can comprehend the meaning of the questions and to be able to spot the correct multiple-choice answer from the 4 possible answers. These questions are specifically designed to give the new Technician Class operator useful, practical knowledge so they can begin operating their equipment as soon as their call sign shows up on the FCC database!

So how long will it take you to prepare for your upcoming exam? Probably about 30 days to work through this book and learn the material. Once you're ready, call The W5YI VEC at 800-669-9594 to find an exam location near you.

QUESTIONS REARRANGED FOR SMARTER LEARNING

The first thing you'll notice when you look at how the Element 2 question pool is presented in this book is that we have *completely rearranged the entire Technician Class question pool*. This rearrangement will take you *logically* through each and every one of the 411 Q & A's. The questions are arranged into 20 topic areas in a way to match up questions on similar topics.

For example, we have taken all of the questions in the pool that talk about *where you can operate your ham radio* and grouped them together into one area that allows you to better understand all of the material that relates to this topic. This arrangement of the questions follows a natural learning process beginning with what ham radio is, what radio waves are, how they get from here to there, a peek under the cover of the modern Technician Class ham set, and ends with safety issues to keep you safe and sound in your new amateur radio hobby.

The reorganization of all of the test questions in the pool has been tested and finely-tuned in Gordo's weekend classes. This method of learning WORKS! You will probably cut the amount of study time in half simply by following the question pool from front to back as presented in this book!

Let us assure you that each and every Element 2 Q&A is in this book. A cross-reference of all 411 questions is found on pages 225 to 226 of the book, along with the syllabus used by the QPC to develop this new Element 2 pool.

This book – and Gordo's Technician Class audio course – contain all 411 Technician Class questions, 4 possible answers, the noted correct answer, and description of how the correct answer works into the real world of amateur radio. We also include ***KEY WORDS*** that will help you remember the correct answer and provide you with a fast review of the entire question pool just before you sit for the Tech exam. Be sure to look for our friend Elmer. He has many **HAM HINTS** to share with you, giving special tips and insights into ham radio operations that better illustrate specific Technician Class on-air techniques. At the end of each section of Q&As, we provide Website Resources – web addresses with information on selected "hot topics" that will help you understand the real world of ham radio.

When you visit some of these websites, it may not be immediately apparent why we are suggesting them. Some addresses take you to the sites of local specialty ham clubs that contain lots of information on how to operate on repeaters, or on satellites, or provide educational resources on learning about electronics or antennas or how radios work.

These addresses were current at the time of publication. However, websites move, addresses change, or sites simply go away. So if you find an address that doesn't work, feel free to drop us an e-mail so we can update our book. Here's our e-mail address: masterpubl@aol.com.

WATCH THIS! ON YOU TUBE

Meet Josh Nass, KI6NAZ, who has thousands of viewers to his HAM RADIO CRASH COURSE YouTube weekly shows, and hundreds of archived videos about ham radio in action!

"The HRCC videos provide content and context for all levels of ham radio, emerging wireless technologies, and ham radio on the air," explains Josh. From programming a new handheld radio to building a simple antenna, his videos are classics that help explain and demonstrate the topics in our ***Technician Class*** book!

At the end of each Q&A topic section in our book, you'll see ***Watch This!*** with the HRCC URL and suggested video numbers to watch. This URL will take you directly to the YouTube playlist for the Ham Radio Crash Course videos that will give you a fun look at what awaits you with your new ham radio Technician Class license!

Watch This! **https://www.youtube.com/playlist?list=PL1KAjn5rGhizk5f2_whBLJaOHj59OCl_3**

How to Read the Questions

Using an actual question from page 33, here is a guide to explain what it is you will be studying as you go through all of the Q&As in the book:

Official Q&A

T1A02 Which agency regulates and enforces the rules for the Amateur Radio Service in the United States?

A. FEMA.
B. Homeland Security.
C. The FCC.
D. All of these choices are correct.

Key Words to Remember

The Federal Communications Commission (*FCC*) makes and enforces all Amateur Radio rules in the United States. [97.1] **ANSWER C**

FCC Part 97 Rule Citation

Correct Answer

Topic Areas

Here is a list of the topic areas showing the page where it starts in the book. Again, there is a complete cross reference list of the Q&As in numerical order on pages 225 to 226 in the Appendix, along with the official Question Pool Syllabus.

STUDY SUGGESTIONS

Here are some suggestions to make your learning easier:

1. Read over each multiple-choice answer carefully. Some answers start out looking good, but turn bad during the last few words. If you speed read the answers, you could easily select a wrong answer. Also, while they won't change any words in the answers, they will sometimes scramble the A-B-C-D order.
2. Keep in mind that there is only one question on your test that will come from each group, and track how many groups in each sub-element.
3. Give this book to a friend, and ask him or her to read you the correct answer. You now give the question wording.
4. Mark the heck out of your book! When the pages begin to fall out, you're probably ready for the exam! Take it with you everywhere you go.
5. The book is available on audio CDs, too. So if you'd like to listen to Gordo and Eric read the questions and answers to you while you're driving your car, riding your bike, or laying on the beach, we can do that for you. The CD symbol, disk number, and track number at the beginning of each topic section keys this book to the audio book. You can get the book on audio CD where you purchased this book, or by calling 800-669-9594, or by visiting **www.w5yi.org**.

   CD 1 **TRACK 2**

6. Highlight the keywords one week before the test. Then speed read the brightly highlighted key words twice a day before the exam.

Are you ready to work through the 411 Q & A's? Put a check mark by the easy ones that you may already know the answer for, and put a little circle by any question that needs a little bit more study. Save your highlighting work until a few days before your upcoming test. Work the Q & A's for about 30 minutes at a time. We drop in a little bit of humor to keep you on track; and if you actually need Gordo's live words of encouragement, you can call him on the phone Monday through Thursday, 10:00 a.m. to 4:00 p.m. (California time), 714-549-5000.

THE QUESTION POOL, PLEASE

Okay, this is the big moment – your Technician Class, Element 2, question pool. Don't freak out and get overwhelmed with the prospect of learning 411 Q & A's. You will find that the topic content is repeated many times, so you're really going to breeze through this without any problem!

About Ham Radio

T1A02 Which agency regulates and enforces the rules for the Amateur Radio Service in the United States?

A. FEMA
B. Homeland Security
C. The FCC
D. All these choices are correct

The Federal Communications Commission (***FCC***) makes and enforces all Amateur Radio rules in the United States. [97.1] **ANSWER C.**

T1A01 Which of the following is part of the Basis and Purpose of the Amateur Radio Service?

A. Providing personal radio communications for as many citizens as possible
B. Providing communications for international non-profit organizations
C. Advancing skills in the technical and communication phases of the radio art
D. All these choices are correct

The Amateur Rules and Regulations very specifically state the basis and purpose of Amateur Radio. Part 97.1 of the FCC rules addresses the distinction between Amateur and Commercial radio. As useful and practical as all of the above may seem, only one of the functions listed reflects the long-standing purposes of Amateur Radio: ***"Advancing skills in the technical and communications phases of the radio art."*** [97.1] **ANSWER C.**

Kids working with adults on a ham radio "foxhunt" advancing their skills by locating hidden radio transmitters in all radio services.

T1C01 For which license classes are new licenses currently available from the FCC?

A. Novice, Technician, General, Amateur Extra
B. Technician, Technician Plus, General, Amateur Extra
C. Novice, Technician Plus, General, Advanced
D. Technician, General, Amateur Extra

While privileges from beloved but no-longer-issued classes of Amateur Radio licenses have been grandfathered into current license classes, no new Technician Plus, Novice, or Advanced class licenses are being issued. Today, the FCC issues only three ham licenses — ***Technician, General, and Amateur Extra.*** Incidentally, many new Technicians are unaware of the value of the high frequency (HF) privileges they "inherited" from the prior system. These old Novice privileges allow worldwide HF communications on four valuable shortwave bands. [97.9(a), 97.17(a)] **ANSWER D.**

Shortwave Listening: A SWL Idea

Shortwave Listening (SWLing) has been around almost as long as radio and is a great way to get your feet wet in radio before you even get your license. Decent shortwave radios are inexpensive and plentiful, and with conditions improving, you can listen in on the world with little effort.

For a large part of ham radio's existence, hams got their first taste of long distance radio by listening to the countless shortwave broadcasters, utility stations, air traffic controllers, ships at sea, and mystery spy stations using shortwave receivers. Check out this site: **swling.com/** for a great overview of shortwave listening.

It is most likely that your first HF transceiver will have general coverage receive capacity, so you don't have to feel *too* guilty about buying one before you get your license! You can even listen to W1AW's code practice sessions so you can be ready to dive right into CW the day you get your "ticket"!

T1C10 How soon after passing the examination for your first amateur radio license may you transmit on the amateur radio bands?

A. Immediately on receiving your Certificate of Successful Completion of Examination (CSCE)
B. As soon as your operator/station license grant appears on the ARRL website
C. As soon as your operator/station license grant appears in the FCC's license database
D. As soon as you receive your license in the mail from the FCC

Good news! You will not need to wait very long to begin operating with your first Amateur Radio license. ***As soon as you see your FCC license grant listed in the FCC's on-line database*** on the Internet you are on the air! If your Volunteer Examination team and VE Coordinator file electronically, you should see your license grant within 48 hours of electronic submission. [97.5(a)] **ANSWER C.**

T1A05 What proves that the FCC has issued an operator/primary license grant?

A. A printed copy of the certificate of successful completion of examination
B. An email notification from the NCVEC granting the license
C. The license appears in the FCC ULS database
D. All these choices are correct

Unlike hams of yore, you no longer have to wait on pins and needles for several weeks for your new amateur radio license to arrive by U.S. mail. ***Once your operator/primary station license shows up in the FCC's online ULS database,*** you're good to go! The FCC no longer mails out paper licenses. When you first see your license on the database, print a copy right away. After that one-time opportunity, any future printouts of you license will have the words REFERENCE COPY on it. If you're like most self-respecting hams, you'll want to make a copy of the downloaded license to display on your wall for all your friends to see! [97.7] **ANSWER C.**

UNITED STATES OF AMERICA
FEDERAL COMMUNICATIONS COMMISSION
AMATEUR RADIO LICENSE
KB9SMG
PETER E TROTTER

T1A04 How many operator/primary station license grants may be held by any one person?

A. One
B. No more than two
C. One for each band on which the person plans to operate
D. One for each permanent station location from which the person plans to operate

Your combination operator/primary station license conveys that call sign to you and only you. You may be a trustee for a number of club or special event stations, but ***you may only have one operator/primary station license.*** [97.5(b)(1)] **ANSWER A.**

T1C08 What is the normal term for an FCC-issued amateur radio license?

A. Five years
B. Life
C. Ten years
D. Eight years

Amateur station licenses are granted for a ***term of 10 years.*** You will need to renew your license at that time, and renewing will be done through an on-line ULS application to the FCC. There is no exam required to renew a valid ham radio license. [97.25] **ANSWER C.**

T1C09 What is the grace period for renewal if an amateur license expires?

A. Two years
B. Three years
C. Five years
D. Ten years

You are not allowed to operate during the 2-year grace period; however, ***you keep your privileges for 2 years.*** After that, they are lost for good. So is your call sign. Don't forget to renew so you don't have to take the exam again! [97.21(a)(b)] **ANSWER A.**

T1C11 If your license has expired and is still within the allowable grace period, may you continue to transmit on the amateur radio bands?

A. Yes, for up to two years
B. Yes, as soon as you apply for renewal
C. Yes, for up to one year
D. No, you must wait until the license has been renewed

How do active, licensed amateur radio operators let their licenses expire? They move and don't inform the FCC of their new mailing address. The FCC has the option of canceling your license if any mail to you is returned. You would never know this cancellation happened unless someone looks up your call sign, sees that it is, and tells you. VEC groups may send you a reminder to renew, but if your address with the FCC is incorrect you'll never receive the notice. Regardless of how your license expires, ***you are not permitted to transmit until the FCC license database shows that your license has been renewed.*** No transmitting during the 2-year grace period. [97.21 (b)] **ANSWER D.**

Website Resources

IF YOU'RE LOOKING FOR	THEN VISIT
W5YI Group/VEC	www.w5yi.org
What is ham radio	www.arrl.org/what-is-ham-radio
Main link to FCC info	wireless.fcc.gov/rules.html
FCC site with lots of info	http://wireless.fcc.gov/services/index.htm?job=service_home&id=amateur
CQ Magazine	www.cq-amateur-radio.com
News for hams needing special help	www.handiham.org
Glossary of general definitions & amateur terms	www.icomamerica.com/en/downloads/default.aspx?Category=352
Using the FCC ULS system	https://youtu.be/s9bB5UTGSOs
How to get an FRN number	https://youtu.be/SaHcLG9cLSO

Watch This! **https://www.youtube.com/playlist?list=PL1KAjn5rGhizk5f2_whBLJaOHj59OCI_3 (videos 14 + 347)**

Call Signs

T1F03 When are you required to transmit your assigned call sign?

A. At the beginning of each contact, and every 10 minutes thereafter
B. At least once during each transmission
C. At least every 15 minutes during and at the end of a communication
D. At least every 10 minutes during and at the end of a communication

Give your call sign regularly – ***every 10 minutes and at the end of your transmission***. Remember, even though the law doesn't require that you give your call sign at the beginning of the transmission, it makes good sense to start out with your call sign and name to let others know who you are. You worked hard for your call sign! Don't be afraid to use it often! [97.119(a)] **ANSWER D.**

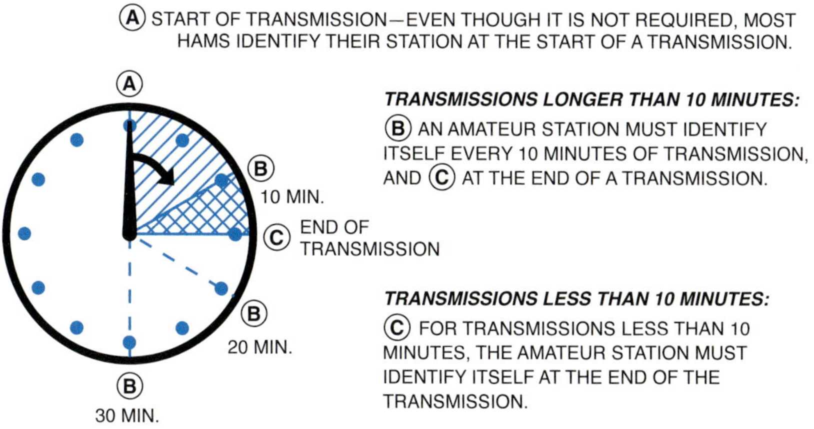

Identifying Amateur Transmissions

T1D11 When may an amateur station transmit without identifying on the air?

A. When the transmissions are of a brief nature to make station adjustments

B. When the transmissions are unmodulated

C. When the transmitted power level is below 1 watt

D. When transmitting signals to control model craft

Unidentified transmissions have always been the big "no-no" of amateur radio. However, recent rules allow for one exception: ***transmissions to operate radio controlled cars, boats, or aircraft***. An alternate means of identifying the transmitter call sign is required, such as physically attaching a call sign label to the remote control transmitter. [97.119(a)] **ANSWER D.**

Hams have several exclusive channels on UHF to control drones. These channels are not shared with the general public, and offer longer range.

T1C02 Who may select a desired call sign under the vanity call sign rules?

A. Only a licensed amateur with a General or Amateur Extra Class license

B. Only a licensed amateur with an Amateur Extra Class license

C. Only a licensed amateur who has been licensed continuously for more than 10 years

D. Any licensed amateur

If you're like most hams, your call sign becomes an important part of your identity! You may not be as fanatical about your call sign as one individual who had his tattooed on his chest, but if you're an active ham, you'll be knit very closely with your "ham ID" for a long time. Far more hams worldwide recognize my call sign than my name. For decades, amateur radio call signs were pretty much the luck of the draw. Now with the Vanity Call Sign program, you can (within limits) pick a call sign that fits your personality. Most radio amateurs prefer short, distinctive call signs but longer call signs can be memorable too. As a Technician Class licensee you may choose your call sign from Group C (1 by 3) or Group D (2 by 3). What does this "1 by 3" and "2 by 3" business mean? A "1 by 3" means one letter before your number (remember all legitimate U.S. call signs have one and only one number), and three letters after. A "2 by 3" call has two letters before the number, and three after. Of course, all vanity calls are assigned on a first come-first-served basis. ***Any licensed amateur can request a vanity call sign.*** Want one? The W5YI-VEC can help you. Call 800-669-9594. [97.19] **ANSWER D.**

T1C05 Which of the following is a valid Technician class call sign format?

A. KF1XXX
B. KA1X
C. W1XX
D. All these choices are correct

Most radio amateurs prefer short, distinctive call signs over longer ones. However, since there are a lot fewer short call signs to go around, these are reserved to higher class licensees. This is a great incentive for many hams to upgrade as soon as possible! As a Technician Class licensee, the vanity call sign you can choose is a slightly longer one, but there are still some distinctive call signs in the 1 x 3 or 2 x 3 categories. KA1X is a 2 x 1 , and W1XX is a 1 x 2 format for higher classes license grades. ***FCC first-issue technician class call signs will always be 2 x 3, two letters, a number, and 3 letters.*** **ANSWER A.**

T1A03 What do the FCC rules state regarding the use of a phonetic alphabet for station identification in the Amateur Radio Service?

A. It is required when transmitting emergency messages
B. It is encouraged
C. It is required when in contact with foreign stations
D. All these choices are correct

The ***use of a phonetic alphabet is*** as old as voice (phone) radio, and is always ***encouraged*** when radio communications are less than ideal. The use of phonetics is not the place to be "creative." Always use the standard ITU phonetic alphabet, shown here, to make your message clear. [97.119(b)(2)] **ANSWER B.**

ITU International Phonetic Alphabet

A – Alpha – AL fah	J – Juliet – JEW lee ett	S – Sierra – SEE air rah
B – Bravo – BRAH voh	K – Kilo – KEY loh	T – Tango – TANG go
C – Charlie – CHAR lee	L – Lima – LEE mah	U – Uniform – YOU nee form
D – Delta – DELL tah	M – Mike – MIKE	V – Victor – VICK tah
E – Echo – ECK ohh	N – November – NO vem ber	W – Whiskey – WISS key
F – Foxtrot – FOKS trot	O – Oscar – OSS car	X – X-ray – ECKS ray
G – Golf – GOLF	P – Papa – PAH pah	Y – Yankee – YANG kee
H – Hotel – HOH tel	Q – Quebec – KEH beck	Z – Zulu – ZOO loo
I – India – IN dee ah	R – Romeo – ROW me o	

T1F11 Which of the following is a requirement for the issuance of a club station license grant?

A. The trustee must have an Amateur Extra Class operator license grant
B. The club must have at least four members
C. The club must be registered with the American Radio Relay League
D. All these choices are correct

Club stations are wonderful tools for increasing and focusing interest in amateur radio. There are countless special interest club stations, such as contesting stations, DXing stations, experimental stations, and more. One of the real benefits of a club station is that you can get a very unique and meaningful call sign, something that helps define and focus your club's purpose. It's not hard to obtain a club call sign; all you need is a club! ***Your club needs to have four members*** and needs to have regular meetings. It's usually not too hard to find four hams that agree on something. [97.5(b)(2)] **ANSWER B.**

T1F02 How often must you identify with your FCC-assigned call sign when using tactical call signs such as "Race Headquarters"?

A. Never, the tactical call is sufficient
B. Once during every hour
C. At the end of each communication and every ten minutes during a communication
D. At the end of every transmission

You will be the only one in the world with your unique amateur radio call sign. When operating on the air, it's always a good idea to give your call sign phonetically. If you're operating a ham radio station at a big event, you are permitted to use tactical call signs, such as "Checkpoint Charlie." However, the rules still require that you ***give your own call sign every ten minutes and at the end of each transmission*** along with your tactical call sign. [97.119(a)] **ANSWER C.**

T1F04 What language may you use for identification when operating in a phone sub-band?

A. Any language recognized by the United Nations
B. Any language recognized by the ITU
C. English
D. English, French, or Spanish

Today is an exciting adventure on the 10 meter band. As a Technician Class operator your new privileges include voice emissions from 28.300 to 28.500 MHz. When the skip goes long, it's possible to hook up with a station in a foreign country. You are permitted to speak their language as a courtesy to the other operator. However, you must give your own call sign every ten minutes in English. Even though you have been speaking fluent Italian for the last 9 minutes and 59 seconds, it's time to ***identify in*** our own language – ***English***. [97.119(b)(2)] **ANSWER C.**

Testing your radio? Give your call sign! In English!

T1F06 Which of the following self-assigned indicators are acceptable when using a phone transmission?

A. KL7CC stroke W3
B. KL7CC slant W3
C. KL7CC slash W3
D. All these choices are correct

Your Alaska station (call sign area 7) is now operating mobile in Maryland, the area 3 call sign district. To let other operators know you are in the area, use your regular call sign followed by the word stroke, or the word slant, or digital indicator of forward slash mark and then "W3." ***All of these choices are correct***. This way, locals won't think Alaska propagation is coming in on the local 2 meter band! They know you're in their neighborhood. [97.119(c)] **ANSWER D.**

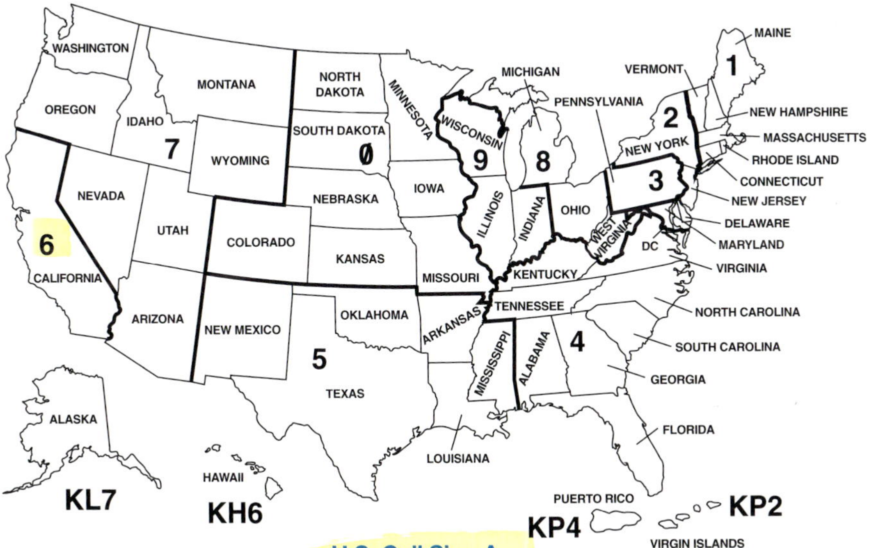

U.S. Call Sign Areas

The number in your new call sign is determined by your permanent mailing address. If you're in Vermont, your call sign might be KB1DOG. In Texas? You're in "5-land." Wisconsin is in call sign district 9. When you move, you may keep your call sign even if you relocate to another number area.

T1D01 With which countries are FCC-licensed amateur radio stations prohibited from exchanging communications?

A. Any country whose administration has notified the International Telecommunication Union (ITU) that it objects to such communications

B. Any country whose administration has notified the American Radio Relay League (ARRL) that it objects to such communications

C. Any country banned from such communications by the International Amateur Radio Union (IARU)

D. Any country banned from making such communications by the American Radio Relay League (ARRL)

Your new Technician Class license gives you many opportunities to talk around the world. A few hams own the distinction of "Worked All Countries" – a little more than 350 throughout the world. The ITU publishes the "Status of Radiocommunications Between Amateur Stations of Different Countries." As of the last printing, there were only two ***countries that prohibit communications between amateur stations***: the Democratic People's Republic of Korea (North Korea), and Eritrea. [97.111(a)(1)] **ANSWER A.**

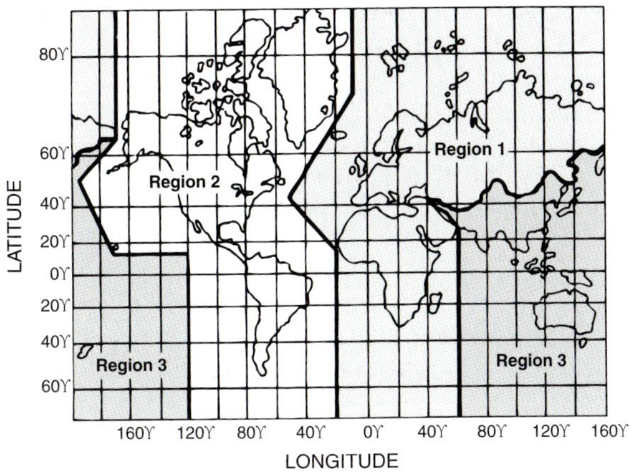

ITU Regions

T1C03 What types of international communications are an FCC-licensed amateur radio station permitted to make?

A. Communications incidental to the purposes of the Amateur Radio Service and remarks of a personal character
B. Communications incidental to conducting business or remarks of a personal nature
C. Only communications incidental to contest exchanges; all other communications are prohibited
D. Any communications that would be permitted by an international broadcast station

Ham operators are not allowed to conduct business over the air nor broadcast messages to shortwave listeners. And ham radio used only for contesting is not what being a good ham operator is about. While a good percentage of your foreign contacts will occur during a contest, it is extremely gratifying to have actual conversations with foreign hams during more relaxed conditions, and these are certainly legal. They're a big part of ham radio! Just be careful to avoid controversial or sensitive topics. Our ***ham transmissions*** are of a ***personal nature***. [97.117] **ANSWER A.**

T1F08 What is the definition of third party communications?

A. A message from a control operator to another amateur station control operator on behalf of another person
B. Amateur radio communications where three stations are in communications with one another
C. Operation when the transmitting equipment is licensed to a person other than the control operator
D. Temporary authorization for an unlicensed person to transmit on the amateur bands for technical experiments

Third-party traffic is any ***communication by amateur radio on behalf of a non-licensed person***. This can include an unlicensed person talking over an amateur radio station, or an amateur passing a radiogram on behalf of a third party. Many foreign countries consider any such communications to be competition

List of Countries Permitting Third-Party Traffic

Country	Call Sign Prefix	Country	Call Sign Prefix	Country	Call Sign Prefix
Antigua and Barbuda	V2	El Salvador	YS	Paraguay	ZP
Argentina	LU	The Gambia	C5	Peru	OA
Australia	VK	Ghana	9G	Philippines	DU
Austria, Vienna	4U1VIC	Grenada	J3	Pitcairn Island	VR6
Belize	V3	Guatemala	TG	St. Kitts & Nevis	V4
Bolivia	CP	Guyana	8R	St. Lucia	J6
Bosnia-Herzegovina	T9	Haiti	HH	St. Vincent & Grenadines	J8
Brazil	PY	Honduras	HR	Sierra Leone	9L
Canada	VE, VO, VY	Israel	4X	South Africa	ZS
Chile	CE	Jamaica	6Y	Swaziland	3D6
Colombia	HK	Jordan	JY	Trinidad and Tobago	9Y
Comoros	D6	Liberia	EL	Turkey	TA
Costa Rica	TI	Marshall Is	V6	United Kingdom	GB
Cuba	CO	Mexico	XE	Uruguay	CX
Dominica	J7	Micronesia	V6	Venezuela	YV
Dominican Republic	HI	Nicaragua	YN	ITU-Geneva	4U1ITU
Ecuador	HC	Panama	HP	VIC-Vienna	4U1VIC

for their commercial radio communications and forbid such traffic. A Third Party Agreement must be in place in order for a radio amateur to participate in this activity. The list of countries allowing third party traffic is on the FCC website. Never assume a foreign country has a third-party agreement. [97.3(a)(47)] **ANSWER A.**

T1F07 Which of the following restrictions apply when a non-licensed person is allowed to speak to a foreign station using a station under the control of a licensed amateur operator?

A. The person must be a U.S. citizen
B. The foreign station must be in a country with which the U.S. has a third party agreement
C. The licensed control operator must do the station identification
D. All these choices are correct

A ***Third Party Agreement*** must be in place in order for a radio amateur to participate in this activity. Never assume a foreign country has a third-party agreement. Always check! [97.115(a)(2)] **ANSWER B.**

T1C06 From which of the following locations may an FCC-licensed amateur station transmit?

A. From within any country that belongs to the International Telecommunication Union
B. From within any country that is a member of the United Nations
C. From anywhere within International Telecommunication Union (ITU) Regions 2 and 3
D. From any vessel or craft located in international waters and documented or registered in the United States

When you get your new ham license, you are permitted to use your call sign anywhere our Federal Communications Commission regulates radio traffic, such as throughout the United States and its territories. We do have agreements in place for permission to operate in several other countries, and we'll learn more

Countries Holding U.S. Reciprocal Agreements

Antigua, Barbuda	Colombia	Grenada	Macedonia	Solomon Islands
Argentina	Costa Rica	Guatemala	Marshall Is.	South Africa
Australia	Croatia	Guyana	Mexico	Spain
Austria	Cyprus	Haiti	Micronesia	St. Lucia
Bahamas	Denmark	Honduras	Monaco	St. Vincent and Grenadines
Barbados	Dominica	Iceland	Netherlands	Surinam
Belgium	Dominican Rep.	India	Neth. Antilles	Sweden
Belize	Ecuador	Indonesia	New Zealand	Switzerland
Bolivia	El Salvador	Ireland	Nicaragua	Thailand
Bosnia-Herzegovina	Fiji	Israel	Norway	Trinidad, Tobago
Botswana	Finland	Italy	Panama	Turkey
Brazil	France[2]	Jamaica	Paraguay	Tuvalu
Canada[1]	Germany	Japan	Papua New Guinea	United Kingdom[3]
Chile	Greece	Jordan	Peru	Uruguay
	Greenland	Kenya	Philippines	Venezuela
		Kiribati	Portugal	
		Kuwait	Seychelles	
		Liberia	Sierra Leone	
		Luxembourg		

1. Do not need reciprocal permit
2. Includes all French Territories
3. Includes all British Territories

about this in up-coming pages of this book. But what happens when you are out sailing on the high seas, in international waters? ***If that sailboat is registered or documented in the U.S.***, seen flying our stars and stripes on the stern flag staff, then you are good to go! You still must abide by the rules for your grade of license, and all the ***FCC rules are still in force***, even though you may be 1000 miles out at sea. Be sure to get the permission of the ship's Captain before going on the air with your little ham radio set. [97.5(a)(2)] **ANSWER D.**

Website Resources

▼ IF YOU'RE LOOKING FOR	▼ THEN VISIT
Call sign look-up	www.qrz.com
Look up a pal's call sign	www.hamcall.net
Available vanity calls	www.radioqth.net/vanity/available
Vanity call sign services	www.w5yi.org

Watch This! **https://www.youtube.com/playlist?list=PL1KAjn5rGhizk5f2_whBLJaOHj59OCI_3 (videos 161)**

Control

T1E01 When may an amateur station transmit without a control operator?

A. When using automatic control, such as in the case of a repeater
B. When the station licensee is away and another licensed amateur is using the station
C. When the transmitting station is an auxiliary station
D. Never

When a ham radio station goes on the air and begins transmitting, there must always be a control operator who is responsible for the transmissions. The control operator might be present at the controls, may control the station remotely, or may have equipment tied into the transmitter to provide automatic control. But ***all transmitting stations must have a control operator*** responsible for the station when it transmits over the airwaves! [97.7(a)] **ANSWER D.**

Chip, K7JA, serves as the control operator on 10 meters during Field Day exercises.

T1E02 Who may be the control operator of a station communicating through an amateur satellite or space station?

A. Only an Amateur Extra Class operator
B. A General class or higher licensee with a satellite operator certification
C. Only an Amateur Extra Class operator who is also an AMSAT member
D. Any amateur allowed to transmit on the satellite uplink frequency

Any amateur can talk over the satellites. When you pass your new Technician Class exam, you'll be allowed to transmit on the satellite uplink frequencies that are within your Technician Class privileges. You'll instantly join the mainstream of satellite activity! [97.301, 97.207(c)] **ANSWER D.**

T1E03 Who must designate the station control operator?

A. The station licensee
B. The FCC
C. The frequency coordinator
D. Any licensed operator

The control operator is the station licensee or another licensed ham designated by the station licensee to be the control operator. It is up to ***the station licensee*** to either control his or her own station or select an alternate control operator who is appropriately licensed. [97.103(b)] **ANSWER A.**

T1E07 When the control operator is not the station licensee, who is responsible for the proper operation of the station?

A. All licensed amateurs who are present at the operation
B. Only the station licensee
C. Only the control operator
D. The control operator and the station licensee

Your Extra Class Elmer comes over to your station and asks to use your 10 meter gear up on 29.600 MHz, FM, in an area that is outside your Technician Class band privileges. This is okay, but your Elmer must give his own call sign along with your call sign, and ***both of you are responsible*** for the transmissions since the gear is yours. Both of you are responsible! A situation like this is fairly common in a contest where a "higher class" contester may be borrowing your station. [97.103(a)] **ANSWER D.**

When you operate by yourself from another ham's station, you use your license class privileges.

T1E04 What determines the transmitting frequency privileges of an amateur station?

A. The frequency authorized by the frequency coordinator
B. The frequencies printed on the license grant
C. The highest class of operator license held by anyone on the premises
D. The class of operator license held by the control operator

Soon you will be a Technician Class operator. Good for you! Did you know you have high frequency SSB voice privileges on the 10 meter band, 28.300 to 28.500 MHz? Your friend who holds an Extra Class license enters the station and wants to run your station on 29.600 MHz FM. If you ***designate your friend as the control operator*** of your station, ***his or her privileges*** with the Extra Class ticket would allow operation well above your normal Technician Class band limits. [97.103(b)] **ANSWER D.**

T1E06 When, under normal circumstances, may a Technician class licensee be the control operator of a station operating in an Amateur Extra Class band segment?

A. At no time
B. When designated as the control operator by an Amateur Extra Class licensee
C. As part of a multi-operator contest team
D. When using a club station whose trustee holds an Amateur Extra Class license

As a Technician Class operator, you get some exciting privileges on worldwide high frequency bands down on 10 meters for voice, and on 15, 40, and 80 meters for CW. You really love Morse code, so could you tune down to the Extra Class portion of the band and run with the experts? No, you must ***stay within your Technician Class band privileges***. [97.301] **ANSWER A.**

Morse Code for Fun

It's been a while since Morse Code proficiency was required for any class of Amateur Radio license, but that's no reason you shouldn't learn it. There are still tens of thousands of radio amateurs who regularly use Morse Code (or CW), for the sheer enjoyment of doing so. Morse Code is now a carrot rather than a stick.

As a Technician, you have three HF bands where you can only use CW, and these are bands with amazing worldwide communications. Even if you upgrade to General or Extra, there is the wonderful 30-meter band that is limited to CW and some digital modes only. (40-meters and 30-meters are my favorite bands).

For weak signal work at the bottom end of our VHF and UHF bands, CW is the way to go. It's also the dominant mode for exotic modes like Moonbounce (EME) and Meteor Burst. And, after you upgrade, and get access to our two new lowest frequency bands on 630 and 2200 meters, you'll find that slow speed CW reigns supreme.

Additionally, if you want to get on HF "on the cheap," nothing beats CW. CW transmitters are very inexpensive, and there are lots of high-performance, low-powered CW rigs in kit form out there, which is a great way to learn electronics and have a blast at the same time.

Someone once said, "A billion people speak Chinese; how hard can it be to learn?" The same can be said for CW. For most of the existence of Amateur Radio every ham knew Morse Code, believe it or not!

There are countless resources for learning Morse Code quickly and painlessly, on-line and elsewhere. No excuses!

T1E05 What is an amateur station's control point?

A. The location of the station's transmitting antenna
B. The location of the station's transmitting apparatus
C. The location at which the control operator function is performed
D. The mailing address of the station licensee

This is ***the spot where you have complete capabilities*** to turn the equipment on or shut it off in case of a malfunction. Every ham radio station is required to have a control point. [97.3(a)(14)] **ANSWER C.**

The control point for ham radio is inside your car when the equipment is installed in your vehicle. Make sure to abide by all "no-distraction" laws.

T1E08 Which of the following is an example of automatic control?

A. Repeater operation
B. Controlling a station over the internet
C. Using a computer or other device to send CW automatically
D. Using a computer or other device to identify automatically

About the only time you will see a big rack of ham radio equipment with no actual operator on duty is the ***automatically controlled*** station. This could be a ***repeater*** up on a mountain top, or a radio link atop a high rise. The FCC rules require that all the details indicating who is in charge be posted on the equipment in case of a malfunction during automatic control. [97.3(a)(6), 97.205(d)] **ANSWER A.**

T1E10 Which of the following is an example of remote control as defined in Part 97?

A. Repeater operation
B. Operating the station over the internet
C. Controlling a model aircraft, boat, or car by amateur radio
D. All these choices are correct

"Anyone on the repeater want to talk to Antarctica? Here they come!" announced the repeater control operator. He's tying into IRLP (Internet Radio Linking Project) using the keypad on his handheld to activate the distant repeater on the Antarctica node using ***remote control***. [97.3(a)(39)] **ANSWER B.**

T1E09 Which of the following are required for remote control operation?

A. The control operator must be at the control point
B. A control operator is required at all times
C. The control operator must indirectly manipulate the controls
D. All these choices are correct

While more and more amateur radio operation is under either remote or automatic control, a live, human control operator must always be present at the helm. Many Technicians simply assume that a repeater station, for instance, just sits up on a hill someplace and runs itself. This is not the case; a licensed control operator must be available to terminate repeater operation if it is functioning in an illegal manner. Contact information for the control operator must be posted in writing at the transmitter location. While the control operator does not need to be at the physical transmitter (not many hams would volunteer to live at the top of a hill with a repeater 24/7), there is always a control operator available at the remote control console for that repeater. The remote control console may be something as simple as a handheld radio with a key pad, or a more elaborate, permanent "control room" arrangement. Just be aware that when you're using a repeater, someone is watching! ***All the choices are correct***. [97.109(c)] **ANSWER D.**

T1E11 Who does the FCC presume to be the control operator of an amateur station, unless documentation to the contrary is in the station records?

A. The station custodian
B. The third party participant
C. The person operating the station equipment
D. The station licensee

Any time you let another licensee run your radio gear, always make an entry into your log book to note who the control operator was at the time. If ***the FCC*** is monitoring, they ***will assume that you were the control operator***, unless you can provide documentation to the contrary in your station records. Always keep a log book, especially when other hams are using your gear or when passing third party traffic. [97.103(a)] **ANSWER D.**

T1D08 In which of the following circumstances may the control operator of an amateur station receive compensation for operating that station?

A. When the communication is related to the sale of amateur equipment by the control operator's employer
B. When the communication is incidental to classroom instruction at an educational institution
C. When the communication is made to obtain emergency information for a local broadcast station
D. All these choices are correct

No transmitting messages on behalf of your employer, and no transmitting that your prize poodle just had a litter with each puppy going for $175. We may not be compensated by any local weather service to take weather reports far out at sea. In other words, no pay for transmitting over ham radio. There is an exception for ***ham radio operators who are school teachers***. They can be "on the clock" during classroom hours demonstrating ham radio for their students to learn what amateur radio is all about. [97.113(a)(3)(iii)] **ANSWER B.**

T1F10 Who is accountable should a repeater inadvertently retransmit communications that violate the FCC rules?

A. The control operator of the originating station.
B. The control operator of the repeater.
C. The owner of the repeater.
D. Both the originating station and the repeater owner.

This question deals with the ham who inadvertently goes into a brain-fade while operating on a repeater. You are late going to work and on the air talking through the local repeater lamenting the terrible traffic congestion. A co-worker ham says he's already in the parking lot and will tell the troops you are going to be late. Without thinking, you ask him to tell your secretary to postpone the sales meeting until tomorrow. ***ERROR!!*** As the sales manager for that company, you are conducting a conversation over the air that applies to your direct business in violation of FCC rules. ***You're responsible.*** [97.205(g)] **ANSWER A.**

Website Resources

▼ IF YOU'RE LOOKING FOR	▼ THEN VISIT
Control Op duties	www.arrl.org/regulatory-faqs
Operating aids and references	www.ac6v.com
San Bernardino Microwave Society	www.ham-radio.com/sbms
Questions about the amateur service rules? Be informed!	www.w3beinformed.org
FCC Amateur Enforcement Actions	http://transition.fcc.gov/eb/AmateurActions
Report rule violations	fccham@fcc.gov
FCC Rules	www.gpo.gov/fdsys [title 47] then [Part 97] and also see Parts 0, 1, 2, 17,and 214.
Purchase a printed copy of Part 97	www.w5yi.org
USA ITU info	www.itu.int
Entire Question Pools	www.ncvec.org

Watch This! **https://www.youtube.com/playlist?list=PL1KAjn5rGhizk5f2_whBLJaOHj59OCI_3 (videos 67 + 84)**

CD 1 TRACK 4

Mind the Rules

T1D06 What, if any, are the restrictions concerning transmission of language that may be considered indecent or obscene?

A. The FCC maintains a list of words that are not permitted to be used on amateur frequencies
B. Any such language is prohibited
C. The ITU maintains a list of words that are not permitted to be used on amateur frequencies
D. There is no such prohibition

You will hear most everything and anything over the ham radio airwaves. But one thing ***absolutely prohibited*** on any ham radio frequency ***is indecent and obscene language***. And while a swear word here and there might get through, rank discussions with foul language are not tolerated. [97.113(a)(4)] **ANSWER B.**

T1A11 When is willful interference to other amateur radio stations permitted?

A. To stop another amateur station that is breaking the FCC rules
B. At no time
C. When making short test transmissions
D. At any time, stations in the Amateur Radio Service are not protected from willful interference

Radio Amateurs share radio spectrum with countless other radio amateurs and other duly licensed radio services. Accidental interference is unavoidable, but ***intentionally interfering*** with other amateur radio communications ***is never acceptable***. Our long-standing reputation as self-policing radio operators depends on all radio amateurs behaving in a civilized manner. [97.101 (d)] **ANSWER B.**

T1D10 How does the FCC define broadcasting for the Amateur Radio Service?

A. Two-way transmissions by amateur stations
B. Any transmission made by the licensed station
C. Transmission of messages directed only to amateur operators
D. Transmissions intended for reception by the general public

You may not operate your station like an AM, FM or shortwave broadcast station ***transmitting directly to the public.*** [97.3(a)(10)] **ANSWER D.**

T1D02 Under which of the following circumstances are one-way transmissions by an amateur station prohibited?

A. In all circumstances
B. Broadcasting
C. International Morse Code Practice
D. Telecommand or transmissions of telemetry

The FCC defines broadcasting as one-way transmission of radio signals intended for a general (non-ham) audience. ***Amateur Radio must not compete with any commercial radio service like commercial broadcasting.*** The transmission of music is prohibited for the same reason. [97.113(b), 97.111(b) **ANSWER B.**

T1D09 When may amateur stations transmit information in support of broadcasting, program production, or news gathering, assuming no other means is available?

A. When such communications are directly related to the immediate safety of human life or protection of property
B. When broadcasting communications to or from the space shuttle
C. Where noncommercial programming is gathered and supplied exclusively to the National Public Radio network
D. Never

Ham radio signals get through when other radio services and cell phones might not. When a big hurricane moves across the southeast, ham radio stations generally stay on the air. Normally, we are not permitted to assist in news gathering for commercial television stations; however, in an emergency, where regular commercial radio and TV broadcast station transmitters have been blown over, ***ham operators are permitted to assist in emergency news gathering***, assuming there is absolutely no other radio system available, and ***assuming that their reports will*** assist in ***protecting lives and property***. [97.113(5)(b)] **ANSWER A.**

T1D03 When is it permissible to transmit messages encoded to obscure their meaning?

A. Only during contests
B. Only when transmitting certain approved digital codes
C. Only when transmitting control commands to space stations or radio control craft
D. Never

This question asks about secret codes and ciphers – intentional scrambling to prevent eavesdropping. This is only ***allowed for those control operators*** who are ***transmitting*** special ***telecommands to ham radio satellites***. Also, hams who enjoy flying their ***model aircraft*** using digital code encryption are permitted to do so to ensure their aircraft doesn't accidentally get the wrong command from another transmitter. [97.211(b), 97.215(b), 97.113(a)(4)] **ANSWER C.**

Ham Hint: *We are beginning to see some "leaks" in proper ham radio distribution, with equipment coming into the U.S. directly from overseas, intended for the land-mobile market, but which work quite nicely on ham frequencies, too. This cheap gear may include voice encryption, and this is not permitted in the ham radio service! So steer clear of imported, no-name ham gear, and never use any type of voice encryption on the ham bands. Check to make sure the equipment you are buying is FCC Certified before handing over your hard-earned cash.*

T1D04 Under what conditions is an amateur station authorized to transmit music using a phone emission?

A. When incidental to an authorized retransmission of manned spacecraft communications
B. When the music produces no spurious emissions
C. When transmissions are limited to less than three minutes per hour
D. When the music is transmitted above 1280 MHz

Music is generally not allowed on the ham bands. No playing the violin or piano, and no singing happy birthday. However, a little-known rule could permit you to blow your trumpet for reveille when sending an authorized signal to the International Space Station. This music is considered ***"incidental to an authorized retransmission of manned spacecraft communications."*** You do play the trumpet, right? [97.113(a)(4), 97.113(c)] **ANSWER A.**

T1D05 When may amateur radio operators use their stations to notify other amateurs of the availability of equipment for sale or trade?

A. Never
B. When the equipment is not the personal property of either the station licensee, or the control operator, or their close relatives
C. When no profit is made on the sale
D. When selling amateur radio equipment and not on a regular basis

Once a week you tune into an interesting, fun radio net where hams, in turn, give their call sign and list a piece of ham radio equipment they own that's up for sale. The ***occasional sale of your own ham radio equipment is not considered a business rule violation*** as long as you don't buy and sell ham radio equipment regularly for profit. So these "swap nets" are perfectly okay, as long as your offering of equipment is on an occasional basis. [97.113(a)(3)(ii)] **ANSWER D.**

T1F01 When must the station and its records be available for FCC inspection?

A. At any time ten days after notification by the FCC of such an inspection
B. At any time upon request by an FCC representative
C. At any time after written notification by the FCC of such inspection
D. Only when presented with a valid warrant by an FCC official or government agent

While it rarely happens to ham operators sticking to the rules, an ***FCC knock on the front door requires you to let them in for an inspection*** of your station equipment and station records. It usually doesn't happen out of the blue – if you are on the air and regularly wiping out TV reception for the entire block, the FCC may contact you and ask you to investigate what is going on. Who knows – it might be a leaky cable problem out your back door. But if the problem persists, the FCC could ask to inspect your station. They will not take "no" for an answer! [97.103(c)] **ANSWER B.**

T1C04 What may happen if the FCC is unable to reach you by email?

A. Fine and suspension of operator license
B. Revocation of the station license or suspension of the operator license
C. Revocation of access to the license record in the FCC system
D. Nothing; there is no such requirement

The FCC needs to be able to contact you at any time, for as long as you are licensed, and email is how they do it. You are responsible to keep your email address current with the FCC. If they can't promptly locate you, ***they could revoke the station license or suspend the operator license***. You must keep the FCC updated on your current e-mail address, as they seldom send postal mail anymore. [97.23] **ANSWER B.**

T1C07 Which of the following can result in revocation of the station license or suspension of the operator license?

A. Failure to inform the FCC of any changes in the amateur station following performance of an RF safety environmental evaluation
B. Failure to provide and maintain a correct email address with the FCC
C. Failure to obtain FCC type acceptance prior to using a home-built transmitter
D. Failure to have a copy of your license available at your station

The FCC needs to be able to contact you if any questions arise pertinent to your station operation. The accepted means of them keeping in touch with you is by email, so ***it is crucial that you keep your email address current with the FCC***. Failure to do so could result in revocation of your station license or suspension of your operator privileges. [97.23] **ANSWER B.**

Watch This! **https://www.youtube.com/playlist?list=PL1KAjn5rGhizk5f2_whBLJaOHj59OCl_3 (videos 346 + 119)**

CD 1 TRACK 5

Tech Frequencies

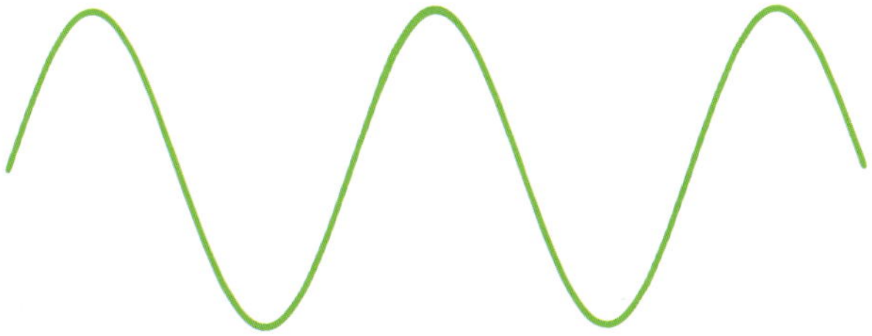

T5C06 What does the abbreviation "RF" mean?

A. Radio frequency signals of all types
B. The resonant frequency of a tuned circuit
C. The real frequency transmitted as opposed to the apparent frequency
D. Reflective force in antenna transmission lines

The term ***"RF"*** refers to radio frequency. If a unit transmits, it puts out a radio frequency. A wireless hot spot has a radio frequency. Your cell phone has a radio frequency output. Your new Technician Class transceiver has a ***radio frequency*** output as soon as you push the transmit button. **ANSWER A.**

T3B03 What are the two components of a radio wave?

A. Impedance and reactance
B. Voltage and current
C. Electric and magnetic fields
D. Ionizing and non-ionizing radiation

The ***electric and magnetic*** fields of a radio wave are at right angles to each other, and together they are called "electromagnetic" radio waves. **ANSWER C.**

T3B01 What is the relationship between the electric and magnetic fields of an electromagnetic wave?

A. They travel at different speeds
B. They are in parallel
C. They revolve in opposite directions
D. They are at right angles

In a radio frequency electromagnetic wave, the magnetic and electrical fields are at right angles to each other. Electrical fields follow the physical polarization of the antenna, usually vertical polarized for mobile and handheld FM and data transmissions on VHF and UHF. **ANSWER D.**

T3B04 What is the velocity of a radio wave traveling through free space?

A. Speed of light
B. Speed of sound
C. Speed inversely proportional to its wavelength
D. Speed that increases as the frequency increases

The velocity of a ***radio wave*** is the same as the ***speed of light***: 300,000,000 meters per second. 300 million meters per second is the same velocity as the speed of light in a vacuum. Oh yeah, radio waves slow down a bit as they go through the ionosphere, clouds, and our smog layers, or travel along the surface of the Earth. This slowing down is very slight but contributes to a large number of very important radio behaviors you will observe, like knife edge diffraction. **ANSWER A.**

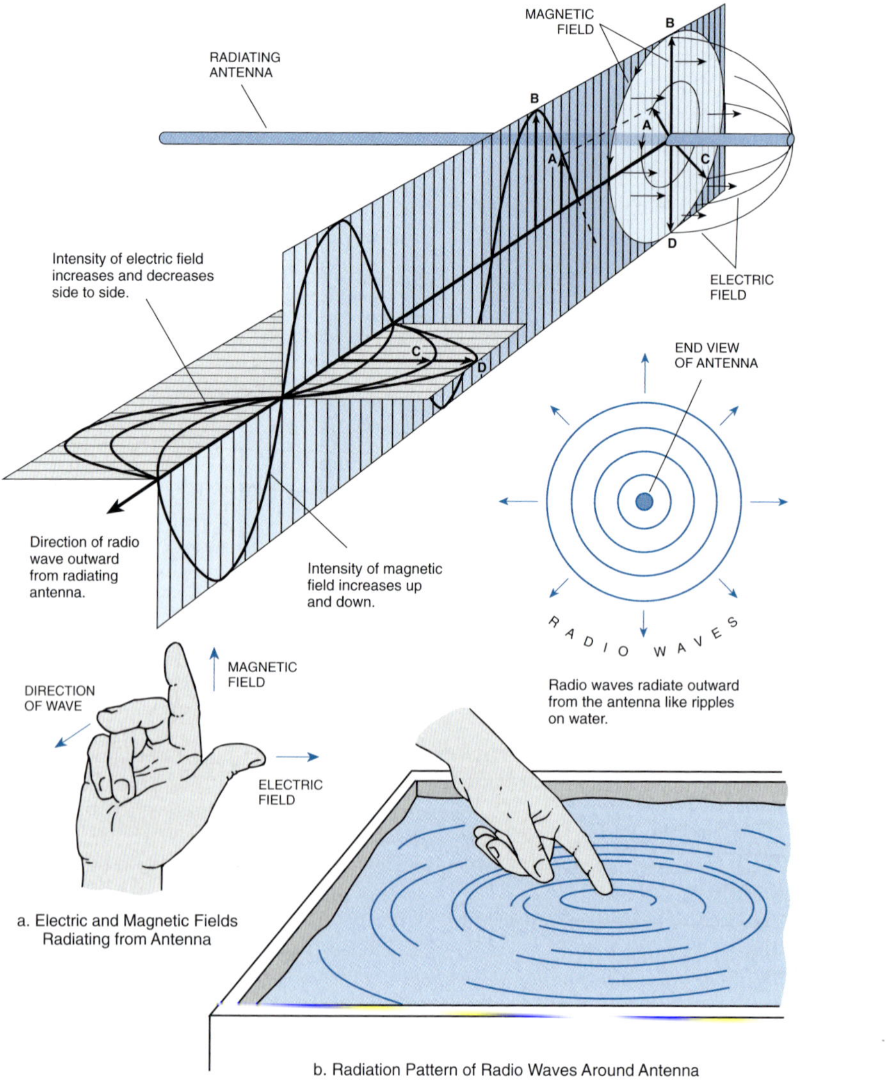

a. Electric and Magnetic Fields Radiating from Antenna

b. Radiation Pattern of Radio Waves Around Antenna

Radio Waves

Source: Basic Electronics © 1994, 2000 Master Publishing, Inc., Niles, Illinois

T3B11 What is the approximate velocity of a radio wave in free space?
A. 150,000 meters per second
B. 300,000,000 meters per second
C. 300,000,000 miles per hour
D. 150,000 miles per hour

The velocity of radio waves through free space is ***300,000,000 meters per second*** – 300 hundred million meters per second. Meters per second, not miles per hour. **ANSWER B.**

T5A12 What describes the number of times per second that an alternating current makes a complete cycle?
A. Pulse rate
B. Speed
C. Wavelength
D. Frequency

When we measure ***frequency***, we count the ***number of times per second that current flows back and forth*** for a complete cycle. And do you remember the definition of the word frequency? Cycles per second, which we officially call hertz, abbreviated Hz. Don't forget the capital H! **ANSWER D.**

T5A06 What is the unit of frequency?
A. Hertz
B. Henry
C. Farad
D. Tesla

The basic unit of frequency is the ***hertz***, abbreviated Hz. **ANSWER A.**

T5C13 What is the abbreviation for kilohertz?
A. KHZ
B. khz
C. kHZ
D. kHz

It's a good idea to memorize all the metric prefixes, both the big ones like kilo and mega, and the little ones like milli and micro. Unfortunately, there's not a lot of consistency when it comes to capitalization of these prefixes. mHz is millihertz, a small unit, beginning with a small m, which makes sense. Megahertz is MHz, which is a large unit, beginning with a capital M, which also makes sense. However, kilohertz, which is a large ***unit has the abbreviation kHz*** with a small k, which makes no sense. So you just need to memorize these things. **ANSWER D.**

T5C07 What is the abbreviation for megahertz?
A. MH
B. mh
C. Mhz
D. MHz

Megahertz means a million hertz, or a million cycles per second. It is correctly abbreviated ***MHz***. Capital M, Capital H, lower case z. **ANSWER D.**

T5B07 Which is equal to 3.525 MHz?
A. 0.003525 kHz
B. 35.25 kHz
C. 3525 kHz
D. 3,525,000 kHz

Move the decimal point 3 places to the right to convert MHz (millions) to kHz (thousands). ***3.525 MHz is 3525 kHz***. Kilohertz means 1,000 cycles per second. Kilohertz is abbreviated kHz – lower case k, Capital H, lower case z. Most HF radios display frequency in kilohertz. **ANSWER C.**

T3B07 In addition to frequency, which of the following is used to identify amateur radio bands?

A. The approximate wavelength in meters
B. Traditional letter/number designators
C. Channel numbers
D. All these choices are correct

When we want to meet a pal on a specific ham band, we usually say which band it is in meters. The most popular ham band for Tech operators is the 2-meter band. Your pal says that she has a 2-meter radio, so the next question is what exact frequency? It's really a tossup on whether you want to tell someone to meet you at a specific frequency or ask them if they have a 2-meter radio, which describes the wavelength of the band on which you plan to operate. If you asked me, I would first determine what ***wavelength band*** we were going to operate on, and then indicate the specific frequency we should try to make contact on. **ANSWER A.**

T3B05 What is the relationship between wavelength and frequency?

A. Wavelength gets longer as frequency increases
B. Wavelength gets shorter as frequency increases
C. Wavelength and frequency are unrelated
D. Wavelength and frequency increase as path length increases

The ***higher*** we go in ***frequency***, the ***shorter*** the ***distance*** between each wave. The lower we go in frequency, the longer the distance between each wave. Now say this out loud – LOWER LONGER, HIGHER SHORTER. Got it? Now look around and see who is staring at you wondering what you're talking about! **ANSWER B.**

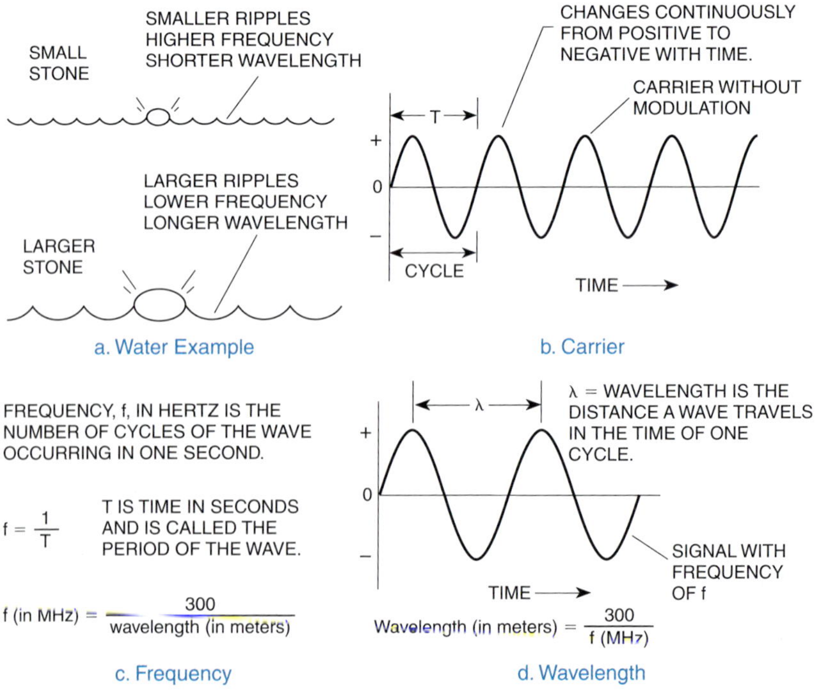

Carrier, Frequency, Cycle and Wavelength

T3B06 What is the formula for converting frequency to approximate wavelength in meters?

A. Wavelength in meters equals frequency in hertz multiplied by 300
B. Wavelength in meters equals frequency in hertz divided by 300
C. Wavelength in meters equals frequency in megahertz divided by 300
D. Wavelength in meters equals 300 divided by frequency in megahertz

Frequency in MHz and wavelength in meters are inversely proportional. The root number to remember is 300! Wavelength is the distance a wave travels in the time of one cycle. Wavelength usually is stated in meters or centimeters. Frequency in hertz is the number of cycles of the wave occurring in one second.

f (in MHz) = 300 ÷ wavelength (in meters)

Wavelength (in meters) = 300 ÷ f (MHz)

On your upcoming examination, the above 4 choices on the test will be identical to these, and they read somewhat similar. Just remember: ***300 ÷ frequency in MHz.*** This correct answer is the only one that ends with the word "megahertz." **ANSWER D.**

Conversions Between Wavelength and Frequency

Converting Frequency to Wavelength

To find wavelength (λ) in meters, if you know frequency (f) in megahertz (MHz), Solve:

$$\lambda(\text{meters}) = \frac{300}{f(\text{MHz})}$$

Converting Wavelength to Frequency

To find frequency (f) in megahertz (MHz), if you know wavelength (λ) in meters, Solve:

$$f(\text{MHz}) = \frac{300}{\lambda(\text{meters})}$$

T3B10 What frequency range is referred to as HF?

A. 300 to 3000 MHz
B. 30 to 300 MHz
C. 3 to 30 MHz
D. 300 to 3000 kHz

High frequency, called ***HF, extends from 3 to 30 MHz***. As a new Technician Class operator, you have privileges on 4 bands of high frequency operation. HF offers exciting daytime and nighttime sky wave "skip" conditions. There is no longer a Morse code test required to operate on the Technician Class high frequency sub-bands. You will need to know Morse code to take advantage of your Morse-code-only privileges on HF 80-, 40-, 15- and 10-meter bands. But good news – on 10 meters, you have single sideband (SSB) voice privileges from 28.3 to 28.5 MHz. During the summertime we get plenty of daytime and late afternoon skip, letting you yak with fellow operators all over the country! We are now at the beginning of Solar Cycle 25, and it looks like a whopper. Sporadic E contacts are independent of solar cycles, and summertime E skip will still give us plenty of DX to all of the U.S. and sometimes multiple hops to Europe, South America, Alaska, and Asia! HF propagation is highly dependent on the 11-year solar cycle. While we are entering Solar Cycle 25, there is a lot of great HF excitement in store. Even during "dead" times, sporadic E can give surprising long-distance communications on 6 and 10 meters. As a brand new Technician Class operator, you may own and operate a large, high-frequency ham radio station, including a major-sized directional antenna system, too! **ANSWER C.**

T1B11 What is the maximum peak envelope power output for Technician class operators in their HF band segments?

A. 200 watts
B. 100 watts
C. 50 watts
D. 10 watts

The typical 3-30 MHz HF transceiver has a nominal output power capability of around 100 watts PEP (Peak Envelope Power), so you are not likely to exceed your licensed power limit using any typical transceiver without an external linear amplifier. However, you are still responsible for determining your ***PEP*** to be sure you are within the ***200-watt limit***. In any case, you should always follow the "prime directive" of amateur radio operation, which says: "An amateur station must use the minimum transmitter power necessary to carry out the desired communications." [97.313] **ANSWER A.**

PEP in Your Step

There are two questions in the Technician Class question pool that deal with Peak Envelope Power (PEP): T1B11, and T1B12. "What does PEP actually mean?," you may ask. It's a very good question, and fortunately, it has a simple answer.

In any *amplitude modulated* radio signal, and that includes AM, SSB, or even CW, the power is continually changing. The radiated power changes at a syllabic rate for phone, or at a character rate for CW. (Some digital modes also use amplitude modulation *as well as* frequency or phase modulation). With these complex, rapidly changing wave patterns, *where and when* exactly do we measure the radio frequency power? The relatively recent PEP rules make it simple. PEP is the absolute *maximum* power encountered at any time during the transmission. It is THIS power, and this power alone, that the FCC is interested in when determining your legal power output.

In a single sideband signal, your PEP can commonly be up to six times the *average* power of your signal (occasionally much higher). Unfortunately, most electromechanical meter movements only follow the AVERAGE power, so you can still greatly exceed the FCC power limits, even though your SWR/POWER meter "needle" may show you being well below those limits.

There are two ways to be sure you are always within the PEP power limits. One is to use an oscilloscope (station monitor) to look at the instantaneous radio frequency peaks, (the much more preferable and more traditional method), or to use a *peak reading* SWR/power meter, a relatively recent innovation. However, even a good peak reading meter needs occasional calibration against... you guessed it... an oscilloscope. So it's best that you learn how to read an oscilloscope right off the bat, if you plan on working much SSB or AM.

T3B08 What frequency range is referred to as VHF?

A. 30 kHz to 300 kHz
B. 30 MHz to 300 MHz
C. 300 kHz to 3000 kHz
D. 300 MHz to 3000 MHz

The ***VHF*** spectrum extends from ***30 to 300 MHz***. Watch out for answer A – this incorrectly lists kHz, not MHz! **ANSWER B.**

T3B09 What frequency range is referred to as UHF?

A. 30 to 300 kHz
B. 30 to 300 MHz
C. 300 to 3000 kHz
D. 300 to 3000 MHz

The ***UHF*** spectrum extends from ***300 to 3000 MHz***. Watch out for answer C – this incorrectly lists kHz, not MHz! **ANSWER D.**

Tech Frequencies

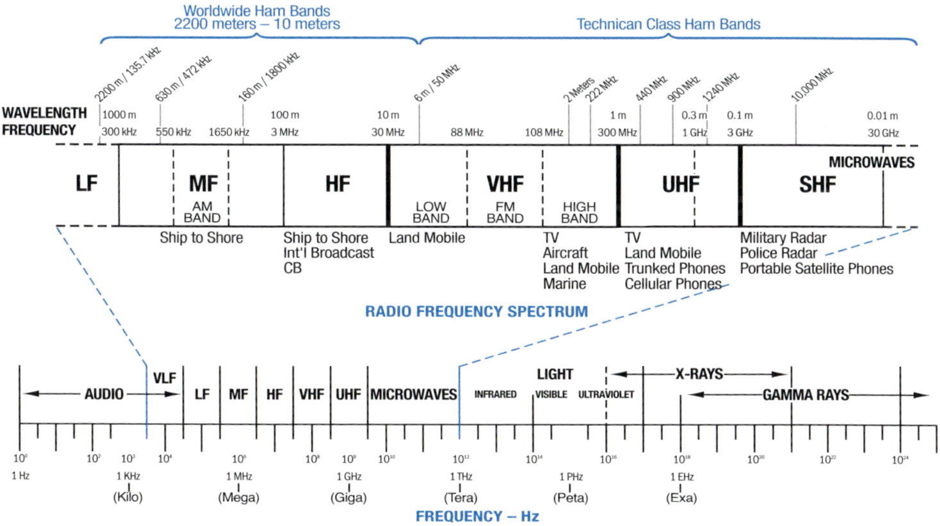

As frequency increases wavelength becomes shorter, as you can see from this RF spectrum chart. Transmissions that require greater bandwidth, such as TV, use higher frequencies.

T1B12 Except for some specific restrictions, what is the maximum peak envelope power output for Technician class operators using frequencies above 30 MHz?

A. 50 watts
B. 100 watts
C. 500 watts
D. 1500 watts

Technician class licensees are permitted 1500 watts of peak envelope power above 30 MHz, and a maximum 200 watts PEP below 30 MHz. Sooner or later, every radio amateur yearns to go QRO (high power). While you want to use your maximum power privileges wisely, there is no reason to avoid using legal limit power when the situation calls for it. Modes like moonbounce, troposcatter, meteor burst and the like can sometimes use all the power you can legally muster! Learning how to work with high powered radio equipment not only makes you more useful during emergencies and various other amateur radio situations, but also can be a launching pad into careers such as broadcast engineering, where high power RF experts are becoming hard to find! ***1500 watts PEP is the absolute maximum allowable power for any class of amateur operation,*** and there are some additional restrictions which may apply. Keep in mind that this 1500-watt limit only applies to transmitter output power, and if you are using a high gain antenna (easy to do above 30 MHz) your effective radiated power (ERP) may greatly exceed 1500 watts and is perfectly legal. [97.313(b)] **ANSWER D.**

T1B03 Which frequency is in the 6 meter amateur band?

A. 49.00 MHz
B. 52.525 MHz
C. 28.50 MHz
D. 222.15 MHz

52.525 MHz is right in the middle of our 6-meter band, which extends from 50 to 54 MHz. Remember f (in MHz) = 300 ÷ wavelength (in meters). 300 ÷ 6 is 50, so your ANSWER is the one that is about 50 MHz. [97.301(a)] **ANSWER B.**

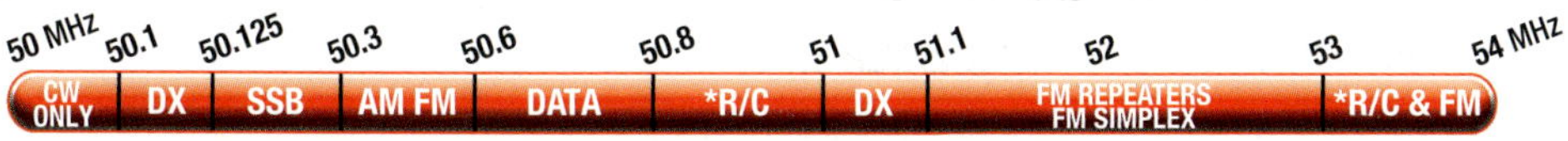

As Easy as PIE

The most important quantity in all of radio (or all of electronics, for that matter) is POWER. It is power, more than anything else, that actually gets things done. And there are few FCC rules related to power that you need to know. Power is easy to calculate; in fact, it's as easy as PIE: P=I x E, where Power (P) is in watts, current (I) is in amperes, and voltage (E) is in volts.

A lot of electronics principles are easier to understand if you connect them with something mechanical. Most of us are familiar with *horsepower.* In fact, there are exactly 746 watts per horsepower. So you can directly translate between electrical power and mechanical power.

It can be somewhat difficult to *measure* power directly, especially Radio Frequency (RF) power, so we usually calculate it using things which *are* easy to measure, like voltage and current.

For measuring DC or low frequency AC power, all you need is a simple digital multimeter, (DMM), which every ham should have in his shack. If you want to measure the power that some device is using, first measure the voltage across the device, while it is in operation.

Next, disconnect one lead from the device in question, and put your DMM in SERIES with the device and the source of power. Place the DMM in the AMPERES mode. (CAUTION: Never place your DMM across any voltage source directly while in the AMPERES mode; you will almost certainly blow the meter's protection fuse, which can be somewhat expensive. The DMM must always be IN SERIES with the device you're testing). Now power up the device and measure the current on the DMM. Multiply this by the voltage you measured in the first step, and that is the power of your device.

As simple as PIE!

T1B04 Which amateur band includes 146.52 MHz?

A. 6 meters
B. 20 meters
C. 70 centimeters
D. 2 meters

146.52 MHz is right in the middle of our ***2-meter band***, which extends from 144 to 148 MHz. Let's do the math: 300 ÷ 146.52 = 2.047502 meters, with 2 meters being the closest answer. Remember, the band refers to an approximate wavelength. Incorrect answers will always be a selection from a band that is *really wrong!* [97.301(a)] **ANSWER D.**

T2A02 What is the national calling frequency for FM simplex operations in the 2 meter band?

A. 146.520 MHz
B. 145.000 MHz
C. 432.100 MHz
D. 446.000 MHz

The national FM simplex calling frequency on the 2 meter band is ***146.520 MHz simplex***, no minus or plus sign on your radio. **ANSWER A.**

T1B05 How may amateurs use the 219 to 220 MHz segment of 1.25 meter band?

A. Spread spectrum only
B. Fast-scan television only
C. Emergency traffic only
D. Fixed digital message forwarding systems only

Use it or lose it! In years past, our 1.25-meter band was 5 MHz wide, extending from 220 to 225 MHz. Due to perceived inactivity by the FCC on our use of the lower portion of the band, we lost 220-222 MHz. But we did recover 1 MHz of prime "radio real estate" at ***219 to 220 MHz, used for point-to-point digital message forwarding***. This is the "backbone" of many digital wireless links in most areas of the country, except in the Mississippi River region, where river navigation traffic has priority use of this sub-band. [97.305(c)] **ANSWER D.**

T2A10 What is a band plan, beyond the privileges established by the FCC?

A. A voluntary guideline for using different modes or activities within an amateur band
B. A list of operating schedules
C. A list of available net frequencies
D. A plan devised by a club to indicate frequency band usage

Every amateur radio band is sliced up into specific operating band plans. ***Band plans are voluntary guidelines*** that hams have established to designate frequencies for specific data signals, voice operation, automatic position reporting system operation, weak signal work, DXing, slow-scan amateur television, propagation beacons, and specific areas for working satellites and the International Space Station. **ANSWER A.**

T1B08 How are US amateurs restricted in segments of bands where the Amateur Radio Service is secondary?

A. U.S. amateurs may find non-amateur stations in those segments, and must avoid interfering with them
B. U.S. amateurs must give foreign amateur stations priority in those segments
C. International communications are not permitted in those segments
D. Digital transmissions are not permitted in those segments

We share the 900-MHz band with the vehicle locator service, which is the primary user of the frequencies. We share 70 cm with the Air Force radio navigation and RADAR service, and they are primary! We must ***avoid interference with primary users***. They have first rights to these frequencies. [97.303] **ANSWER A.**

T1B09 Why should you not set your transmit frequency to be exactly at the edge of an amateur band or sub-band?

A. To allow for calibration error in the transmitter frequency display
B. So that modulation sidebands do not extend beyond the band edge
C. To allow for transmitter frequency drift
D. All these choices are correct

You wouldn't sit on a roof ledge 100 stories up, would you? Same thing with ham radio operation – we don't operate on the edge of a ham radio band. Since none of our radio emissions are allowed beyond the ham band upper or lower limit, we ***stay well within band edges*** in case our signals should ***drift***, or some of our signal ***extends beyond the band edge***, perhaps because our radio is not properly calibrated. Never operate right on the edge of a ham band! [97.101(a), 97.301(a-e)] **ANSWER D.**

Website Resources

▼ IF YOU'RE LOOKING FOR	▼ THEN VISIT
Ham Band Plans and Operating Frequencies	www.ac6v.com/frequencies.htm
Sooner or later you're going to be using a computer with your amateur radio stationHamsphere has a lot of features for the Technician Class operator as well as more advanced operators.	http://www.hamsphere.com/ham-radio.html
Here is a recent buyer's guide for handheld radios.	https://www.twowayradiotalk.com/best-handheld-ham-radios/

Here is a listing of the websites of the major manufacturers of amateur radios sold in the U.S. You can visit their websites to learn more about their handheld, mobile, and base station radios. The manufacturers do not sell directly to consumers, so you will need to shop for your radio at one of the dealers you can find listed on the manufacturer's website.

Alinco handheld and mobile radios	www.alinco.com/usa.html
ICOM America, full line of radios	www.icomamerica.com/en/amateur
Kenwood, full line of radios	www.kenwoodusa.com/Support/Amateur_Radio/
Yaesu, full line of radios	http://yaesu.com/
Wouxun, value-priced handhelds	www.wouxun.com

Watch This! **https://www.youtube.com/playlist?list=PL1KAjn5rGhizk5f2_whBLJaOHj59OCl_3 (video 130)**

Your First Radio

T7A07 What is the function of a transceiver's PTT input?

A. Input for a key used to send CW
B. Switches transceiver from receive to transmit when grounded
C. Provides a transmit tuning tone when grounded
D. Input for a preamplifier tuning tone

A transceiver's PTT (Push To Talk) switch (usually part of a microphone) switches the transceiver from receive to transmit. Many contest operators prefer an STT (Stomp To Talk) foot switch, however, as it allows them to free up their hands for more important tasks, like typing in log data and spinning VFO dials. **ANSWER B.**

The PTT button switches your radio from receive to transmit.

T4B04 What is a way to enable quick access to a favorite frequency or channel on your transceiver?

A. Enable the frequency offset
B. Store it in a memory channel
C. Enable the VOX
D. Use the scan mode to select the desired frequency

There are literally hundreds of hot frequencies in most major cities in the U.S. As a new ham, looking at a frequency guide alone won't necessarily give you a clue as to which ones are great stations to join in on conversations. We recommend that you purchase your handheld through an authorized ham radio dealer that might store about ten of the hot local channels in the memory of your radio for you. Mail order companies may do this, too, as long as they know where the action is on the radio dial in your city. ***Storing popular frequencies in a memory channel*** will allow you to quickly go from one repeater to another, ready to transmit if the conversation sounds inviting! **ANSWER B.**

With a transceiver like one of these, you can hold your ham station in the palm of your hand.

T9A04 What is a disadvantage of the short, flexible antenna supplied with most handheld radio transceivers, compared to a full-sized quarter-wave antenna?

A. It has low efficiency
B. It transmits only circularly polarized signals
C. It is mechanically fragile
D. All these choices are correct

When you open the box containing your new dual-band handheld, you'll find a small "wall wart" battery charger, the rechargeable battery pack or a battery tray, an instruction manual, and the famous "rubber duck" antenna. This flexible antenna is the bare minimum for getting a good signal out on the airwaves. It's okay for transmitting simplex, direct from you to a buddy a mile away, but for reaching out to a repeater more than five miles away, the stock, factory-supplied, ***low-efficiency rubber duck antenna*** is a compromise and is not as effective as a full-sized antenna. **ANSWER A.**

T9A07 What is a disadvantage of using a handheld VHF transceiver with a flexible antenna inside a vehicle?

A. Signal strength is reduced due to the shielding effect of the vehicle
B. The bandwidth of the antenna will decrease, increasing SWR
C. The SWR might decrease, decreasing the signal strength
D. All these choices are correct

If you attempt to transmit using the rubber duck antenna inside your vehicle, your signal will barely make it out the windows. All that metal will make your ***signal 10 to 20 times weaker***. From inside the vehicle, the ham operator should use an external antenna with that VHF radio. **ANSWER A.**

Modern dual- and tri-band handheld transceivers like these have amazing built-in capabilities that make ham radio easy, fun, and portable!

T7A10 What device increases the transmitted output power from a transceiver?

A. A voltage divider
B. An RF power amplifier
C. An impedance network
D. All these choices are correct

As a new Technician Class operator, you'll probably choose a dual-band handheld as your first radio. You can run this in your vehicle using an outside antenna and achieve great results! We would also suggest a filtered DC adaptor plug and a headset speaker/mic. If you really need more than the five watts of power that comes out of the handheld, you also could purchase ***an RF power amplifier*** that would boost the 5 watts to up to 30 watts power output. But you'll probably find a home or automobile outside antenna may be all you need for successful handheld operation from inside your vehicle or home. Antenna improvements should always be considered before increasing power, as an outside antenna will increase both your transmitting and receiving performance. **ANSWER B.**

T8A04 Which type of modulation is commonly used for VHF and UHF voice repeaters?

A. AM
B. SSB
C. PSK
D. FM or PM

We use ***frequency modulation (FM) and phase modulation (PM)*** on most VHF and UHF repeater systems. **ANSWER D.**

T8A09 What is the approximate bandwidth of a VHF repeater FM voice signal?

A. Less than 500 Hz
B. About 150 kHz
C. Between 10 and 15 kHz
D. Between 50 and 125 kHz

When we operate our 2-meter, 1-1/4-meter, and 70- cm handheld using frequency modulation, ***the FM signal is between 10 and 15 kHz wide***. With properly-adjusted equipment 10 kHz total FM bandwidth is about normal. There are repeater systems that operate narrow-band and you must set your radio to the narrow band mode or you will break up on modulation peaks. Narrow band is typically 5 to 7 kHz wide. If you can't get an older radio to go narrow band, just talk softer and further away from the microphone! **ANSWER C.**

T8A02 What type of modulation is commonly used for VHF packet radio transmissions?

A. FM or PM
B. SSB
C. AM
D. PSK

The majority of ***VHF and UHF*** communications ***use frequency modulation, FM***, or ***phase modulation, PM.*** Since VHF data packet communications are on the VHF band, we use FM or phase modulation for them, too. **ANSWER A.**

Ham Hint: *Ask the radio seller to please preprogram at least ten memory channels before you take delivery of your brand-new, dual-band radio. This way, you'll have some popular repeaters ready to go as soon as you turn it on!*

Watch This! **https://www.youtube.com/playlist?list=PL1KAjn5rGhizk5f2_whBLJaOHj59OCl_3 (videos 9, 37, 117, + 54)**

Website Resources

IF YOU'RE LOOKING FOR	THEN VISIT
Ham Equipment Reviews	EHAM.net
Ham Radio Outlet	hamradio.com
Largest Ham Accessory Catalog	MFJenterprises.com
Associated Radio	associatedradio.com
Ham City / Jun's	hamcity.com
DX Engineering	DXEngineering.com
Electronic components and surplus parts	alaskit.net
GigaParts	gigaparts.com
HomeTek	cheapham.com
KJI Electronics	kjielectronics.com
Main Trading Company	mtcradio.com
R&L Electronics	randl.com
Universal Radio	universal-radio.com
Wired Communications	wiredco.com
Like most hams, you'll probably discovered that the latest batch of imported radios are far better than their documentation. Here you can get real documentation in plain English that covers all the nooks and crannies that are overlooked in the "official" operator manual for the very popular Baofeng UV-5r.	http://radiodoc.github.io/uv-5r/
Here is a Linux operating system developed just for the radio amateur! It's a great resource for the experimental-minded radio amateur.	https://sourceforge.net/projects/kb1oiq-andysham/
Batteries	www.Bioennopower.com

Note: This list includes many ham radio dealers where you can purchase your first radio and accessories

CD 2 TRACK 2

Going Solo – Your First Amateur Radio Transmission

T4B02 Which of the following can be used to enter a transceiver's operating frequency?

A. The keypad or VFO knob
B. The CTCSS or DTMF encoder
C. The Automatic Frequency Control
D. All these choices are correct

Most new Technician Class hams start with a dual-band handheld – the popular bands are 2 meters and 70 centimeters. The radio will have a ***keypad*** as well as a ***variable frequency oscillator*** (VFO) knob or up-and-down push buttons. The keypad, VFO knob, and push buttons are a great way ***to select the frequency*** of choice. **ANSWER A.**

Use the keypad or VFO knob to enter your operating frequency.

T2B13 What is the purpose of a squelch function?

A. Reduce a CW transmitter's key clicks
B. Mute the receiver audio when a signal is not present
C. Eliminate parasitic oscillations in an RF amplifier
D. Reduce interference from impulse noise

When there is no signal coming into an FM receiver, there is a lot of white noise generated by the very high gain of the receiver circuits. Stronger signals will "quiet" this noise; FM signals are usually characterized by the "amount of quieting," with a strong signal being called a "full quieting" signal. ***A squelch "kills" the receiver's white noise when no incoming signal is present*** to quiet it. **ANSWER B.**

T4B03 How is squelch adjusted so that a weak FM signal can be heard?

A. Set the squelch threshold so that receiver output audio is on all the time
B. Turn up the audio level until it overcomes the squelch threshold
C. Turn on the anti-squelch function
D. Enable squelch enhancement

The squelch was basically invented to eliminate the extremely annoying noise on an FM receiver in an "unquieted" (no or low signal) state. ***By "opening up" the squelch to hear the annoying noise, you might potentially hear a weak signal buried in the noise.*** However, FM is *not* a suitable mode for weak signal work. Although there is noise on any weak signal mode, the background noise on an SSB or CW receiver is *far far* less obnoxious than unquieted FM noise. And, FM noise has a frequency distribution that makes very weak signals far less intelligible than SSB or CW. **ANSWER A.**

T2B03 Which of the following describes a linked repeater network?

A. A network of repeaters in which signals received by one repeater are transmitted by all the repeaters in the network
B. A single repeater with more than one receiver
C. Multiple repeaters with the same control operator
D. A system of repeaters linked by APRS

Linked repeater systems can ***connect vast swaths of real estate, often across many states.*** Be aware that if you are using a ***network of linked repeaters you will be keying up multiple repeaters*** every time you key your mike. It is always a good idea to use simplex (See T2A11) whenever possible.
Visit: **www.winsystem.org** **ANSWER A.**

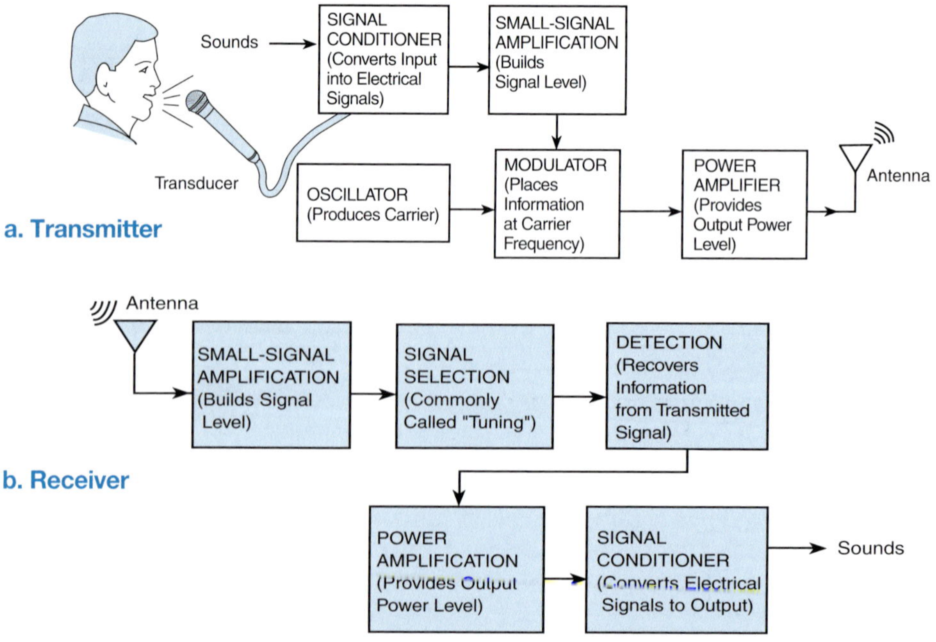

Block Diagram of a Basic Radio Communications System
Source: *Basic Communications Electronics*, Hudson & Luecke,
© 1999, Master Publishing, Inc., Niles, Illinois

T2A11 What term describes an amateur station that is transmitting and receiving on the same frequency?

A. Full duplex
B. Diplex
C. Simplex
D. Multiplex

Simplex means same frequency. You can operate simplex on VHF or UHF when the other station is within a few miles of your station. The opposite of simplex is duplex, a type of repeater operation. A great way to get started with your brand-new call sign is to transmit on 146.520 MHz, simplex. This is the 2-meter national calling channel and a great place to meet new ham friends who are nearby. On 70 cm, try simplex on 446.000 MHz. You'll be surprised by how far your signal may go if you get up on a hill or high up on a building. **ANSWER C.**

T2B09 Why are simplex channels designated in the VHF/UHF band plans?

A. So stations within range of each other can communicate without tying up a repeater
B. For contest operation
C. For working DX only
D. So stations with simple transmitters can access the repeater without automated offset

It is actually good practice to use ***simplex*** frequencies whenever possible. It ***saves congestion on your local repeater***. The reverse split function when working a repeater channel on your handheld is a great way to determine, quite quickly, whether simplex operation is possible. You might be surprised how frequently good, full-quieting simplex operation works! The National Simplex frequency for 2-meters is 146.520 MHz, and many local communities have additional popular simplex "watering holes." In Fairbanks, AK, 52.525 is the 6-meter watering hole, with reliable coverage of most of interior Alaska. **ANSWER A.**

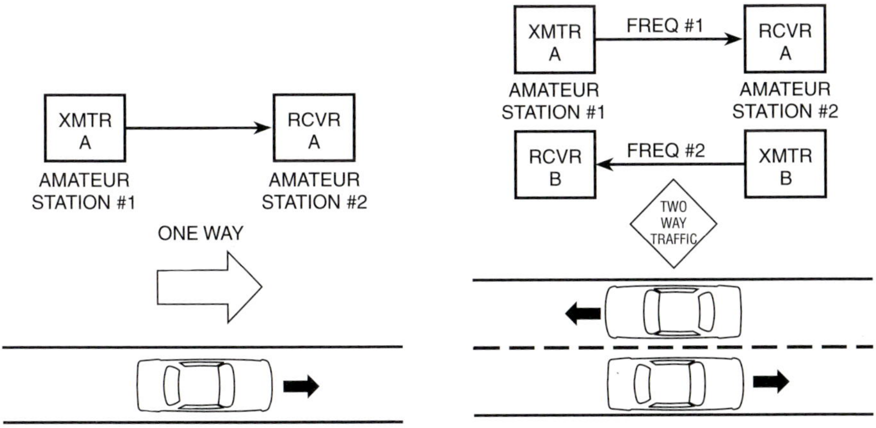

Only one direction at a time – directly from Transmitter to Receiver

a. Simplex (Same Frequency)

Both directions at the same time – from Transmitter A to Receiver A and from Transmitter B to Receiver B

b. Full Duplex (Two Frequencies)

Simplex and Duplex Communications

T2A06 Which of the following is required when making on-the-air test transmissions?

A. Identify the transmitting station
B. Conduct tests only between 10 p.m. and 6 a.m. local time
C. Notify the FCC of the transmissions
D. All these choices are correct

Even a short ***test transmission*** to check out your new microphone, headset, or antenna requires ***station identification***. Try to do your antenna testing on a simplex frequency to avoid tying up a repeater unnecessarily. Many tests, like testing a bad mic cord, can be performed without going on the air at all simply by transmitting into a non-radiating dummy load and monitoring with a local receiver. Any time you can perform a test using a dummy load, do so! A dummy load converts your signal to heat and won't send any signal more than a few hundred feet. **ANSWER A.**

Ham Hint: *Ham radio operators enjoy tinkering with their equipment. If you're at your workbench testing the quality of your transmission by sending a signal into a non-radiating dummy load, there is still plenty of signal for you to monitor with another receiver. Try to avoid any continuous testing when hooked up to an outside antenna, unless you are testing the antenna itself.*

Sooner or later most hams want to "dig into" their radios. Repairing, modifying, and designing new radios are some of the most rewarding aspects of amateur radio. Don't be daunted by what's "under the hood!" All it takes is practice and the guidance of a good Elmer. A well-equipped test bench like the one shown here helps too, but you don't need a lot of fancy equipment right away. Allow your ham shack to grow naturally. And remember, your ham "ticket" is only the beginning of your radio education!

T2A08 What is the meaning of the procedural signal "CQ"?

A. Call on the quarter hour
B. Test transmission, no reply expected
C. Only the called station should transmit
D. Calling any station

The two letters ***"CQ" mean "calling any station,"*** and we use this on all worldwide bands and weak signal calls over VHF and UHF frequencies. But "CQ" is never used when operating on an FM repeater and simplex frequencies because the presence of your carrier is strong enough to let everyone else know you are on the air. Instead of calling "CQ" over a repeater, you simply announce your call sign and indicate you are monitoring for a call. If you're on the air for the very first time, tell them you are a Gordo grad and that may be all that is necessary to bring back plenty of responses from the ham community welcoming you to the exciting airwaves! Try this: "This is (your call sign repeated twice, phonetically) on the air for the first time, a friend of Gordon West, looking for my very first contact on ham radio. Over." This will certainly get attention, and likely you'll get plenty of calls. Always be sure to say your call sign slowly, using the phonetic alphabet. **ANSWER D.**

T2A12 What should you do before calling CQ?

A. Listen first to be sure that no one else is using the frequency
B. Ask if the frequency is in use
C. Make sure you are authorized to use that frequency
D. All these choices are correct

While violating A or B may earn you a reputation as a "lid" (a derogatory term for an inept or impolite operator that dates back to land-line telegraphs days), violating C will earn you a "pink slip" from the FCC. As a self-respecting radio amateur operator, you will want to be both polite and legal. So follow ***all of these*** steps. **ANSWER D.**

T2A05 How should you respond to a station calling CQ?

A. Transmit "CQ" followed by the other station's call sign
B. Transmit your call sign followed by the other station's call sign
C. Transmit the other station's call sign followed by your call sign
D. Transmit a signal report followed by your call sign

What a great day on 6 meters! Signals are coming in via skywaves from over 1,000 miles away. Up at 50.140, there is a young lady calling CQ, and then she stands by for a return call. It's your turn to transmit! Immediately key your microphone, ***say HER call sign once, and then give YOUR call sign phonetically a couple of times***. Release your push-to-talk button, and chances are she will return your call sign and your conversation begins on 6 meters skywave. **ANSWER C.**

T4B12 What is the result of tuning an FM receiver above or below a signal's frequency?

A. Change in audio pitch
B. Sideband inversion
C. Generation of a heterodyne tone
D. Distortion of the signal's audio

Tuning an FM receiver above or below a signal's carrier frequency will result in distortion of the audio. Likewise, ***overmodulating (over-deviating) an FM transmitter will cause distortion in a receiver***; just a small amount of over-deviation can render a receiver's audio totally unintelligible. **ANSWER D**

T2A04 What is an appropriate way to call another station on a repeater if you know the other station's call sign?

A. Say "break, break," then say the station's call sign
B. Say the station's call sign, then identify with your call sign
C. Say "CQ" three times, then the other station's call sign
D. Wait for the station to call CQ, then answer

Before transmitting on any frequency, be sure to listen for a few seconds to ensure the channel is clear. Then depress the microphone push-to-talk button and ***say the call sign of the station you are wishing to hook up with, followed by your call sign***, phonetically, and the optional word "over." Never say "Break Break" to get into a conversation. "Break Break" is only used to signify priority/emergency traffic! **ANSWER B.**

Ham Hint: Getting On The Air!

As you prepare to pass the exam, get set to join a fraternity of fellow ham radio operators. In less than a month, studying one section a day, you will be prepared to find your local examination session and take your test. Visit **www.w5yi.org** *to find an exam site near you, or a remote, on-line session.*

If you get stuck in your studies, call Gordo direct at 714-549-5000 (Monday through Thursday, 10 am to 4 pm California time). He'll be your personal Elmer!

When you pass the exam, ask your VE team for local club information. JOIN A HAM CLUB! The club members will give you lots of help learning the amateur radio ropes. To find a club near you, visit: **www.arrl.org/clubsearch.**

YOUR FIRST RADIO should be a dual-band handheld. This single piece of equipment will get you on the air on your club's local repeater. Most clubs operate repeaters on 2 meters and 70 cm. The 2 meter and 70 cm bands are the most popular repeater bands throughout the world, so get started with a dual-band handheld.

Some good accessories for your 2m/440 MHz handheld are an alkaline battery tray, extra-long rubber antenna, and a combination speaker/microphone. For mobile use, get the 12 volt DC car adapter, light-weight mobile magnetic mount antenna, the mobile antenna adapter, and a hands-free headset for use while driving.

Okay, you got your radio and all the goodies. What's next? Some ham radio sellers may pre-program local repeaters into your radio in the area where you live and work. Purchasing the Repeater Directory will help you find repeaters in your area. Getting a local club member to "clone" or pre-program channels into your handheld can get you going in a hurry.

If you end up with a dual-band handheld with nothing pre-programmed, just remember R-O-O-T. Let's get to the root of the programming function on your handheld.

First, dial in the REPEATER OUTPUT.

Next, check that the OFFSET automatically comes up on the screen, or select the offset MINUS or PLUS.

Use the Repeater Directory to look up the TONE. Tone must first be switched on to ENCODE, and then the specific TONE CODE must be entered. Now try to bring up the repeater by pushing the PUSH-TO-TALK button and listen for the BEEP. Always give your call sign during this transmission. Got BEEP? Now MEMORIZE these details in an open memory channel on your radio.

Trust us, with a little practice you'll get these basics down in no time. You really should know how to do these basics on your radio. If you're stumped, get your radio cloned by a local club member or ask a pal to load in some local hot repeater channels.

T3A01 Why do VHF signal strengths sometimes vary greatly when the antenna is moved only a few feet?

A. The signal path encounters different concentrations of water vapor
B. VHF ionospheric propagation is very sensitive to path length
C. Multipath propagation cancels or reinforces signals
D. All these choices are correct

When a radio signal reflects off a man-made or natural obstacle it can combine with a direct signal to either ***reinforce or partially cancel the resultant signal*** (constructive or destructive interference). The antenna only has to move ½ wavelength to have the two signals switch between constructive and destructive interference. **ANSWER C.**

T3A06 What is the meaning of the term "picket fencing"?

A. Alternating transmissions during a net operation
B. Rapid flutter on mobile signals due to multipath propagation
C. A type of ground system used with vertical antennas
D. Local vs long-distance communications

When mobile stations running either a handheld or a 50-watt mobile to an outside antenna get to the end of their line-of-sight range to a repeater, the signal will begin to ***flutter rapidly***. This is caused by roadway signs and guardrails adding to and subtracting from the signal strength. In the radio business we call this ***"picket fencing."*** A new term for your ham radio vocabulary! **ANSWER B.**

T2B08 Which of the following applies when two stations transmitting on the same frequency interfere with each other?

A. The stations should negotiate continued use of the frequency
B. Both stations should choose another frequency to avoid conflict
C. Interference is inevitable, so no action is required
D. Use subaudible tones so both stations can share the frequency

Common courtesy should prevail on the Amateur Radio bands. Our reputation of being self-policing depends on our long-standing cooperative behavior. Amateur bands are crowded, and occasional, unintentional interference is a fact of life: but it must never be intentional! Making a habit of listening before transmitting, regardless of band or mode, will go a long way toward reducing unintentional interference. While no one has an absolute right to an amateur frequency, ham radio operators should be aware of local net frequencies and times and keep those frequencies clear during those times. Or better yet, participate in the net! After all, we are a community and should be aware of what other hams in our area are doing. Long-standing nets distribute current information and provide an opportunity to practice message passing, a vital amateur radio skill. **ANSWER A.**

T2B10 Which Q signal indicates that you are receiving interference from other stations?

A. QRM
B. QRN
C. QTH
D. QSB

When skywaves bounce off the ionosphere on the 6-meter band, you'll sometimes hear several stations returning your CQ (calling any station) call, all transmitting at the same time. Think of the M in QRM for Man-made interference. This is called ***"QRM," the Q-code for many other stations accidentally interfering*** with each other. **ANSWER A.**

POPULAR Q SIGNALS

Given below are a number of Q signals whose meanings most often need to be expressed with brevity and clarity in amateur work. (Q abbreviations take the form of questions only when each is sent followed by a question mark.)

QRG Will you tell me my exact frequency (or that of _____)? Your exact frequency (or that of _____) is _____ kHz.
QRH Does my frequency vary? Your frequency varies.
QRI How is the tone of my transmission? The tone of your transmission is _____ (1. Good; 2. Variable; 3. Bad).
QRJ Are you receiving me badly? I cannot receive you. Your signals are too weak.
QRK What is the intelligibility of my signals (or those of _____)? The intelligibility of your signals (or those of _____) is _____ (1. Bad; 2. Poor; 3. Fair; 4. Good; 5. Excellent).
QRL Are you busy? I am busy (or I am busy with _____). Please do not interfere.
QRM Is my transmission being interfered with? Your transmission is being interfered with _____ (1. Nil; 2. Slightly; 3. Moderately; 4. Severely; 5. Extremely).
QRN Are you troubled by static? I am troubled by static _____ (1-5 as under QRM).
QRO Shall I increase power? Increase power.
QRP Shall I decrease power? Decrease power.
QRQ Shall I send faster? Send faster (_____ WPM).
QRS Shall I send more slowly? Send more slowly (_____ WPM).
QRT Shall I stop sending? Stop sending.
QRU Have you anything for me? I have nothing for you.
QRV Are you ready? I am ready.
QRW Shall I inform _____ that you are calling on _____ kHz? Please inform _____ that I am calling on _____ kHz.
QRX When will you call me again? I will call you again at _____ hours (on _____ kHz).
QRY What is my turn? Your turn is numbered _____.
QRZ Who is calling me? You are being called by _____ (on _____ kHz).
QSA What is the strength of my signals (or those of _____)? The strength of your signals (or those of _____) is _____ (1. Scarcely perceptible; 2. Weak; 3. Fairly good; 4. Good; 5. Very good).
QSB Are my signals fading? Your signals are fading.
QSD Is my keying defective? Your keying is defective.
QSG Shall I send _____ messages at a time? Send _____ messages at a time.
QSK Can you hear me between your signals and if so can I break in on your transmission? I can hear you between my signals; break in on my transmission.
QSL Can you acknowledge receipt? I am acknowledging receipt.
QSM Shall I repeat the last message which I sent you, or some previous message? Repeat the last message which you sent me [or message(s) number(s) _____].
QSN Did you hear me (or _____) on _____ kHz? I heard you (or _____) on _____ kHz.
QSO Can you communicate with _____ direct or by relay? I can communicate with _____ direct (or by relay through _____).
QSP Will you relay to _____? I will relay to _____.
QST General call preceding a message addressed to all amateurs and ARRL members. This is in effect "CQ ARRL."
QSU Shall I send or reply on this frequency (or on _____ kHz)?
QSW Will you send on this frequency (or on _____ kHz)? I am going to send on this frequency (or on _____ kHz).
QSX Will you listen to _____ on _____ kHz? I am listening to _____ on _____ kHz.
QSY Shall I change to transmission on another frequency? Change to transmission on another frequency (or on _____ kHz).
QSZ Shall I send each word or group more than once? Send each word or group twice (or _____ times).
QTA Shall I cancel message number _____? Cancel message number _____.
QTB Do you agree with my counting of words? I do not agree with your counting of words. I will repeat the first letter or digit of each word or group.
QTC How many messages have you to send? I have messages for you (or for _____).
QTH What is your location? My location is _____.
QTR What is the correct time? The time is _____.

Source: ARRL

T2B11 Which Q signal indicates that you are changing frequency?

A. QRU
B. QSY
C. QSL
D. QRZ

If you find that your station is accidentally causing QRM to an ongoing conversation, it is time to ***QSY – change to another frequency***. Think of the Y in QSY like a Y in the road, to change directions. **ANSWER B.**

T8C03 What operating activity involves contacting as many stations as possible during a specified period?

A. Simulated emergency exercises
B. Net operations
C. Public service events
D. Contesting

While you won't hear a lot of this on your 2-meter/70-centimeter (440) handheld, you will hear this on 6 meters and 2-meter single sideband – contesting. It usually takes place every couple of months over a single weekend. The idea is to contact as many other stations as possible and exchange specific station details, as if you were handling an emergency message. ***Contesting*** is a great way to double-check the performance of your radio system, and it assists you in preparing for an emergency when you may need to contact as many stations as possible. **ANSWER D.**

T8C04 Which of the following is good procedure when contacting another station in a contest?

A. Sign only the last two letters of your call if there are many other stations calling
B. Contact the station twice to be sure that you are in his log
C. Send only the minimum information needed for proper identification and the contest exchange
D. All these choices are correct

During Field Day, as well as on special weekends during the year, our ham bands become more populated with contesters. We try to work as many other stations as we can in many different geographic areas. Many contesters are dead serious about winning, so they don't have time for long-winded pleasantries during the contest exchange. ***Only transmit your call sign and the information they need*** and save chitchat for another weekend! These contesters need to make the exchange fast and move on quickly. Be sure to check the rules ahead of time for the required contest exchange. **ANSWER C.**

During a contest, keep your communications short!

T8C05 What is a grid locator?

A. A letter-number designator assigned to a geographic location
B. A letter-number designator assigned to an azimuth and elevation
C. An instrument for neutralizing a final amplifier
D. An instrument for radio direction finding

When you get active on VHF weak-signal work using a multi-mode radio for single sideband, you'll hear the weak-signal operators saying their ***location as a grid square***. Grid squares are 2-letter by 2-number designators based on 1 degree latitude by 2 degrees longitude. Grid square maps are a handy operating aid when the band opens on 6 meters! My location is DM13, and last night a chap broke in and gave me a signal report from EL95, the tip of Florida! Imagine my surprise, but not that unusual for the 6-meter magic band that you'll hear on the audio course downloaded from **www.w5yi.org.** So check out your grid square map for weak-signal VHF/UHF operating – more than likely, another station is going to ask, "What's your grid?" Go on-line and search for Grid Square Map to see many examples of this useful tool. **ANSWER A.**

Ham Hint: *By the way, there is some kind of contesting going on every weekend to match your particular ham radio interests. Check out the various contest calendars on the web for detailed information. Here's a favorite:* **www.contestcalendar.com/perpetualcal.php**

Website Resources

IF YOU'RE LOOKING FOR	THEN VISIT
Ham clubs near you!	www.arrl.org/find-a-club
Just for Ladies	www.YLRL.org
Making your first contact	www.arrl.org/get-on-the-air
A good portal site for many "absolute beginner" ham radio resources	http://hamradiobeginner.com/
Reducing 2 meter interference	https://youtu.be/FTjiym2KJko

Watch This! **https://www.youtube.com/playlist?list=PL1KAjn5rGhizk5f2_whBLJaOHj59OCl_3 (videos 41 + 76)**

CD 2 TRACK 3

Repeaters

T1F09 What type of amateur station simultaneously retransmits the signal of another amateur station on a different channel or channels?

A. Beacon station
B. Earth station
C. Repeater station
D. Message forwarding station

The device that retransmits amateur radio signals within a specific ham band is called a ***repeater***. When you get your new Technician Class license, first do about one week of monitoring – without transmitting – listening to repeater communications. This will give you a good idea of what the proper operating procedures are for that local repeater frequency. Also, join the local repeater club and let club members help you program your new radio equipment for some of the local repeater frequencies so you know that you have all the details right and can contact the repeater. [97.3(a)(40)] **ANSWER C.**

Here is the heart of a repeater station. On the left is a commercial transmitter and receiver. On the right are duplexer "cans" that allow a single repeater antenna to do double duty. The "cans" are resonant circuits that pass repeater transmit signals to the antenna, while at the same time allowing the repeater receiver to hear the original offset signals coming in from the same antenna system. They also minimize interference to other systems in the same building and on the shared tower.

T1D07 What types of amateur stations can automatically retransmit the signals of other amateur stations?

A. Auxiliary, beacon, or Earth stations
B. Earth, repeater, or space stations
C. Beacon, repeater, or space stations
D. Repeater, auxiliary, or space stations

Up on a mountaintop, ham operators have established ***auxiliary and repeater stations*** that automatically retransmit your signal to other ham stations. Out in space, ***space stations*** incorporate automatic equipment to retransmit the radio signal of ham stations. Tune to 145.825 MHz for the sound of packet from the International Space Station, which you can read on a modern handheld with a built-in GPS and terminal node controller. Imagine sending your call sign and the greeting "Hello World" in data through the International Space Station, and getting a response a few hours later from the other side of the World! [97.113(d)] **ANSWER D.**

T2A07 What is meant by "repeater offset"?

A. The difference between a repeater's transmit and receive frequencies
B. The repeater has a time delay to prevent interference
C. The repeater station identification is done on a separate frequency
D. The number of simultaneous transmit frequencies used by a repeater

A conventional (analog) repeater needs to have its ***transmit and receive signals on different frequencies*** to avoid feedback. In the U.S., 2-meter repeaters typically have their receive and transmit frequencies 600 kilohertz apart, or an offset of 600 kHz. Different offsets are used for different bands. **ANSWER A.**

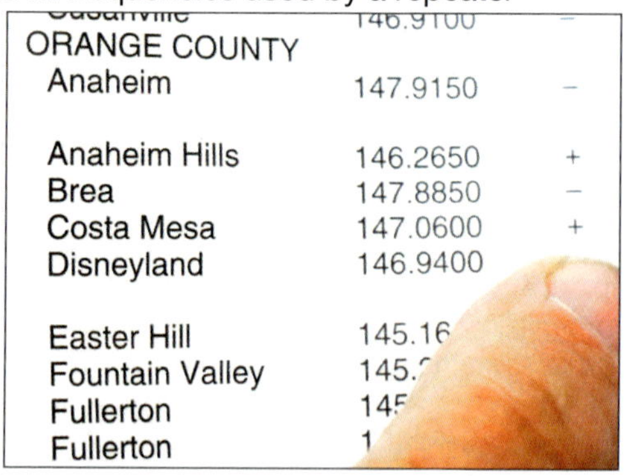

Repeater directories show the repeater offset.

Ham Hint: *Repeater directories publish the repeater frequencies by output. The plus (+) or minus (-) indicates the input "split" that you dial in on your VHF or UHF ham set. A plus (+) indicates a higher input and a minus (-) indicates a lower input. When you start to transmit, your transmitter should automatically go to the proper input frequency. Some repeaters also require a sub-audible tone as part of your input transmission. Ask the local operators how to engage the tone signal on your ham radio set.*

T2B01 How is a VHF/UHF transceiver's "reverse" function used?

A. To reduce power output
B. To increase power output
C. To listen on a repeater's input frequency
D. To listen on a repeater's output frequency

The ***reverse split*** function of your handheld radio allows you to quickly ***monitor the repeater input frequency***. By listening to the repeater input frequency, you might be able to hear direct any user who's transmitting on that repeater's input frequency in your area, allowing you to change to simplex operation. **ANSWER C.**

Repeaters

T2A01 What is a common repeater frequency offset in the 2 meter band?

A. Plus or minus 5 MHz
B. Plus or minus 600 kHz
C. Plus or minus 500 kHz
D. Plus or minus 1 MHz

Most new 2 meter/440 MHz dual-band ham radio equipment has automatic repeater offset capability. This means, on the 2 meter band, your transmitter will ***offset plus or minus 600 kHz*** automatically for you when you tune to the repeater frequency. The offset could go either way, depending on the band plan: plus or minus. On your new radio, the 600 kHz offset may show as .6 MHz on the display. **ANSWER B.**

The display on a handheld usually shows the repeater offset.

T2A03 What is a common repeater frequency offset in the 70 cm band?

A. Plus or minus 5 MHz
B. Plus or minus 600 kHz
C. Plus or minus 500 kHz
D. Plus or minus 1 MHz

Happy to report that modern, dual-band ham radio equipment offers automatic offsets on the 70 cm, 440 MHz, band. Most often we use a ***5 MHz offset, plus or minus***, on the 70 cm band. **ANSWER A.**

T2B04 Which of the following could be the reason you are unable to access a repeater whose output you can hear?

A. Improper transceiver offset
B. You are using the wrong CTCSS tone
C. You are using the wrong DCS code
D. All these choices are correct

You just brought home your new dual-band handheld ham radio but forgot to ask the seller to preprogram some of your local channels. Always ask your dealer if he can preprogram your new transceiver. It will prevent initial disappointment because there's a lot to know at first. Today's frustration is that you've tuned into a local powerful repeater just 1 mile away, but no matter what you do no one hears you when they call you during the "net." ***The repeater may require a tone burst*** that you didn't know about or, more than likely, the repeater requires a specific ***CTCSS tone*** in order for your signal to pass through. Or maybe the repeater requires a ***digital coded squelch tone*** for access. ***All of these*** could spell frustration unless you get your radio preprogrammed. Once you have the right tone and offset for repeater access you'll be on the air! **ANSWER D.**

T2B02 What term describes the use of a sub-audible tone transmitted along with normal voice audio to open the squelch of a receiver?

A. Carrier squelch
B. Tone burst
C. DTMF
D. CTCSS

Every mountain top and skyscraper probably has a few ham repeaters atop a small tower. These repeaters are hearing so many signals coming in, including interfering signals, that they need a way of not accidentally self-triggering and turning on when there isn't a real signal. What makes a signal into a ham repeater real? The repeater may employ ***CTCSS tone*** decode, requiring the ham to encode a specific sub-audible tone that causes the repeater receiver to accept the signal. CTCSS stands for Continuous Tone Coded Squelch System, showing on your radio display as a T for tone. Do not program your radio just yet for CT or else you might not hear anything on channel – just the T, not CT. **ANSWER D.**

Ham Hint: You Need The Tone, Too!

Just as important as the correct repeater split frequency is the correct tone. Tone? Your new dual band handheld radio or mobile radio has a tone encode and decode feature called Continuous Tone Coded Squelch, CTCSS for short, or "tone" for real short. You must select the correct tone for the repeater you have chosen to transmit on. There are 40+ tone possibilities, and the repeater guide will tell you what tone to ENCODE in your radio. A little "t", or "enc" will show up on your screen, indicating you are transmitting a slight "hum" to activate that repeater. The tone will not allow access if the repeater is considered "closed." You need the correct signal to activate the closed repeater system. Your radios have the tone circuit already built in, so it is up to you, or the ham programming your radio, to encode the correct tone for that particular repeater. And if the repeater IS closed, first check with the repeater owner to obtain the correct tone to access the system.

Remember, support your local repeaters, open or closed!

EIA Standard Subaudible CTCSS (PL) Tone Frequencies

Freq.	Tone No.	Tone Code	Freq.	Tone No.	Tone Code	Freq.	Tone No.	Tone Code
67.0	01	XZ	110.9	15	2Z	179.9	29	6B
71.9	02	XA	114.8	16	2A	186.2	30	7Z
74.4	03	WA	118.8	17	2B	192.8	31	7A
77.0	04	XB	123.0	18	3Z	203.5	32	M1
79.7	05	SP	127.3	19	3A	206.5		8Z
82.5	06	YZ	131.8	20	3B	210.7	33	M2
85.4	07	YA	136.5	21	4Z	218.8	34	M3
88.5	08	YB	141.3	22	4A	225.7	35	M4
91.5	09	ZZ	146.2	23	4B	229.2		9Z
94.8	10	ZA	151.4	24	5Z	233.6	36	
97.4	11	ZB	156.7	25	5A	241.8		M5
100.0	12	1Z	162.2	26	5B	250.3		M6
103.5	13	1A	167.9	27	6Z	256.3		M7
107.2	14	1B	173.8	28	6A			

T2A09 Which of the following indicates that a station is listening on a repeater and looking for a contact?

A. "CQ CQ" followed by the repeater's call sign
B. The station's call sign followed by the word "monitoring"
C. The repeater call sign followed by the station's call sign
D. "QSY" followed by your call sign

Simply ***say your call sign*** with a smile in your voice and ***announce that you are monitoring*** for any call. **ANSWER B.**

T1A08 Which of the following entities recommends transmit/receive channels and other parameters for auxiliary and repeater stations?

A. Frequency Spectrum Manager appointed by the FCC
B. Volunteer Frequency Coordinator recognized by local amateurs
C. FCC Regional Field Office
D. International Telecommunication Union

Frequency coordination for VHF and UHF band plans is developed by ***regional frequency coordinators***. It is a huge job because seasoned hams all want their own repeater frequency pairs. Now add requests for simplex coordination for voice-over-Internet systems, and you'll see that frequency coordination in any local area must balance the needs of all ham radio operators. When you purchase your new two-band VHF/UHF handheld, be sure to buy a USA repeater atlas. This way, when you travel to see your family and friends in San Diego, Seattle, Miami, and Connecticut, you'll know what frequencies and what tones to use for the local repeaters. The "locals" will have fun working you on the airwaves as an out-of-town guest, and even though you may be driving in a strange area you'll have plenty of chatter from all your new ham radio friends around you using their local repeater. [97.3(a)(22)] **ANSWER B.**

Ham Hint: *Okay, you have your brand new dual-band handheld, and it's tied into your hidden attic antenna. Your batteries are all charged up, and the local radio dealer memorized a wonderful repeater that you've been listening to all evening long. Now it's time for YOU to make your first transmission.*

First, listen for a couple of minutes to make sure the repeater is not in use. Next, momentarily adjust the squelch so you get background noise, and then adjust the volume to about mid-scale. Now re-adjust the squelch to block background noise. Then press the push-to-talk for approximately 2 seconds, let go, and listen for the repeater to go BEEP. This is the repeater courtesy tone, and most repeaters have a tone that comes on after you release the microphone button to signal the end of that particular transmission. A few repeaters may just show up as a strong single on your handheld signal-strength meter, and then silently click off.

Whether you reached a repeater or not, you MUST give your call sign – so press the push-to-talk button, wait about 1 second, and then clearly state your call letters phonetically: "Kilo zulu six hotel alpha mike, on the air for the first time, brand new ham, listening. Over." Now release the push-to-talk button.

You should hear a beep to confirm your signal was indeed passed through the repeater, and a few seconds later you will probably hear someone calling your call sign and then giving their call sign. If you can, try to write down their call sign.

Now it's your turn to talk. Wait for the beep! You'll press the push-to-talk button, and nothing will come out of your mouth! After all, this is your first transmission, and stage fright is very common. Tell them your first name, where you are located, the fact that this is your first transmission and you're scared to pieces, and maybe find out where the local ham radio club meets because your instructor, Gordo, always said to join a local ham club. Now say over, and release the push-to-talk button.

After that, you'll be rolling with ham radio. Be sure to give your call sign every 10 minutes and when you sign off.

T1A09 Who selects a Frequency Coordinator?

A. The FCC Office of Spectrum Management and Coordination Policy.
B. The local chapter of the Office of National Council of Independent Frequency Coordinators.
C. Amateur operators in a local or regional area whose stations are eligible to be repeater or auxiliary stations.
D. FCC Regional Field Office.

Frequency coordinators are fellow ham operators in a local or regional area, ***voted into their positions by fellow hams who are active repeater station owners*** or auxiliary station operators. Frequency coordination keeps our ham radio frequencies and bands clear of interference. [97.3(a)(22)] **ANSWER C.**

T1F05 What method of call sign identification is required for a station transmitting phone signals?

A. Send the call sign followed by the indicator RPT
B. Send the call sign using a CW or phone emission
C. Send the call sign followed by the indicator R
D. Send the call sign using only a phone emission

Some repeaters have a young lady who greets you with a repeater call sign. Morse code (CW) is always allowed as an identification method, but don't exceed 20-wpm because the FCC says so. [97.119(b)(2)] **ANSWER B.**

Website Resources

IF YOU'RE LOOKING FOR	THEN VISIT
Print versions of ARRL Repeater Directory	www.arrl.org
TravelPlus repeater mapping software	www.arrl.org
Repeater database updated daily	www.artscipub.com/repeater
Software + cables to clone repeater Frequencies from computer to radios	www.rtsystemsinc.com
IRLP info	status.irlp.net
All about IRLP	www.IRLP.net
Echolink info	www.echolink.org
Winlink info	www.winlink.org
ICOM D-Star	www.icomamerica.com/en/products/amateur/dstar/dstar/default.aspx
Yaesu WIRES II Internet radio linking	http://www.yaesu.com/jp/en/wiresinfo-en/index.html

Watch This! **https://www.youtube.com/playlist?list=PL1KAjn5rGhizk5f2_whBLJaOHj59OCl_3 (videos 220, 172, 175, + 325)**

CD 2 TRACK 4

Emergency!!!

T2C06 What is the Amateur Radio Emergency Service (ARES)?

A. A group of licensed amateurs who have voluntarily registered their qualifications and equipment for communications duty in the public service
B. A group of licensed amateurs who are members of the military and who voluntarily agreed to provide message handling services in the case of an emergency
C. A training program that provides licensing courses for those interested in obtaining an amateur license to use during emergencies
D. A training program that certifies amateur operators for membership in the Radio Amateur Civil Emergency Service

ARES is a communications service comprised of hams who volunteer their services. It generally requires certain qualifications beyond just your amateur radio license to fully participate. ARES is often a local service for NGOs (non-government organizations) like the Red Cross and Salvation Army. Many local amateur radio clubs have a strong ARES presence. The ARRL has upgraded its ARES program adding structure, qualifications, and training requirements, and has developed regional data bases of active ARES members in case of a national call up. **ANSWER A.**

Amateur radio operators are well known for their volunteer assistance in emergencies—from local problems to national disasters like 9/11 and Hurricane Katrina.

T2C07 Which of the following is standard practice when you participate in a net?

A. When first responding to the net control station, transmit your call sign, name, and address as in the FCC database
B. Record the time of each of your transmissions
C. Unless you are reporting an emergency, transmit only when directed by the net control station
D. All these choices are correct

Check in just once, and ***don't transmit again until directed to do so by the net control operator.*** **ANSWER C.**

Ham Hint: *During an emergency net, the net control station needs all incoming transmissions to be brief and contain only the information they ask for. Once you pass that information, go into the listen mode. In a widespread emergency, an undisciplined check-in where the ham goes on for 3 minutes to describe all the training he has received over the years in emergency preparedness simply clogs the network.*

T2C01 When do FCC rules NOT apply to the operation of an amateur station?

A. When operating a RACES station
B. When operating under special FEMA rules
C. When operating under special ARES rules
D. FCC rules always apply

Your amateur license is governed by the Federal Communications Commission. You are bound by ***FCC rules***, so any request from the FBI, FEMA, or any other Federal agency does not relieve you from obeying FCC rules. [97.103(a)] **ANSWER D.**

T1A10 What is the Radio Amateur Civil Emergency Service (RACES)?

A. A radio service using amateur frequencies for emergency management or civil defense communications
B. A radio service using amateur stations for emergency management or civil defense communications
C. An emergency service using amateur operators certified by a civil defense organization as being enrolled in that organization
D. All these choices are correct

RACES stands for Radio Amateur Civil Emergency Service. It is a division of the civil defense organization. You must be registered to take part in RACES drills. Notice that ***the three correct answers all contain the RACES key term "civil defense!"*** [97.3(a)(38), 97.407] **ANSWER D.**

T2C04 What is RACES?

A. An emergency organization combining amateur radio and citizens band operators and frequencies
B. An international radio experimentation society
C. A radio contest held in a short period, sometimes called a "sprint"
D. An FCC part 97 amateur radio service for civil defense communications during national emergencies

Over the years there have been a number of Amateur Radio auxiliary emergency services. The two most popular ones nowadays are RACES (Radio Amateur Civil Emergency Service) and ARES (Amateur Radio Emergency Service). A RACES member is trained, certified, and ***enrolled by a civil defense organization***. **ANSWER D.**

In an emergency, authorized hams participating in a RACES organization may communicate from a police helicopter.

T2C09 Are amateur station control operators ever permitted to operate outside the frequency privileges of their license class?

A. No
B. Yes, but only when part of a FEMA emergency plan
C. Yes, but only when part of a RACES emergency plan
D. Yes, but only in situations involving the immediate safety of human life or protection of property

In a very few cases, all of which involve potential "loss of life or limb" conditions, "all bets are off" when it comes to permissible amateur conditions. Be sure if you use this "free pass" you do indeed have a bona fide emergency, and that you are able to document this later on should the FCC (or other official entity) call you on the carpet. This is another reason you should always keep a station log. Most self-respecting hams keep station logs even though they are no longer an FCC requirement. **ANSWER D.**

Ham Hint: *As an ARES or RACES member, you can expect to be on the air at least once or twice a week on your specific net time. Once a month you will attend a local meeting, and regularly you will accrue additional training to make you a better emergency responder. You will have a distinctive uniform, a set callout plan when you are informed that a major emergency has occurred nearby, and a ham radio "grab-and-go" kit that will keep you on the air for at least 48 hours on your assignment. Above all, you must be a REGULAR on the repeater, a REGULAR when it comes to in-person training sessions, and a REGULAR in accruing additional training available to your unit.*

T2C02 Which of the following are typical duties of a Net Control Station?

A. Choose the regular net meeting time and frequency
B. Ensure that all stations checking into the net are properly licensed for operation on the net frequency
C. Call the net to order and direct communications between stations checking in
D. All these choices are correct

One of the oldest activities in Amateur Radio is "round table" or net operation. There are both formal nets and informal nets, but they all work in a similar fashion. The ***net control operator*** is a designated amateur whose job it is to ***call the net to order*** and arrange communications between stations checking in. Net control operators usually operate in a cycle. If you are part of a net, you might be called upon to be a net control operator on occasion and might find yourself being a "regular!" **ANSWER C.**

Ham Hint: *Here's an important exception to the fundamental rule that you can use any radio on any frequency to summon help in an emergency: Avoid contacting a radio service that prohibits any radio call coming in from unknown units. This would include secure military nets, law enforcement radio service, fire radio service, FBI, and other agencies that ONLY communicate among themselves. Using your ham radio for out-of-band transmissions to local police, fire, and state agencies could cost you your ham license when proven you had other radio services, like the U.S. Coast Guard, that you could call who stand by for incoming emergency radio traffic.*

T2C05 What does the term "traffic" refer to in net operation?

A. Messages exchanged by net stations
B. The number of stations checking in and out of a net
C. Operation by mobile or portable stations
D. Requests to activate the net by a served agency

Everyone should know how to take and pass ***formal message traffic***, even if you don't participate in a regular traffic net. When done properly, the formal radiogram is a very reliable means for making sure the message arrives at its destination fully intact. The American Radio Relay League started out as a means of passing reliable third-party traffic, hence the "Relay" in its name. The National Traffic System (NTS) is a highly organized, formal network for passing third party traffic across the continent. If you're interested in public service, it's well worth your while to learn how the NTS works. **ANSWER A.**

T2C10 What information is contained in the preamble of a formal traffic message?

A. The email address of the originating station
B. The address of the intended recipient
C. The telephone number of the addressee
D. Information needed to track the message

Gordo works with fellow ham instructors training members of the military for their Technician Class licenses.

It is important to ***keep track of emergency messages*** as they pass through the well-structured amateur radio traffic-handling system. The make-up of the preamble of the message gives us the details to know where that message came from and where it is going on down the line. **ANSWER D.**

T2C08 Which of the following is a characteristic of good traffic handling?

A. Passing messages exactly as received
B. Making decisions as to whether messages are worthy of relay or delivery
C. Ensuring that any newsworthy messages are relayed to the news media
D. All these choices are correct

A good emergency traffic handler will always print their radio traffic in block letters, word for word. This way, when they ***pass the message on*** to authorities it will be ***exactly as received*** – written word for word as spoken! When using voice, be sure to use standard phonetics if conditions are less than ideal. Learn how to use a standard format RADIOGRAM as well, which is specifically designed to reduce errors. **ANSWER A.**

T2C03 What technique is used to ensure that voice messages containing unusual words are received correctly?

A. Send the words by voice and Morse code
B. Speak very loudly into the microphone
C. Spell the words using a standard phonetic alphabet
D. All these choices are correct

The operative word here is standard. The purpose of phonetics is to make yourself understood under poor conditions. ***Always use*** the ITU ***standard phonetics***, which are recognized worldwide. **ANSWER C.**

T2C11 What is meant by "check" in a radiogram header?

A. The number of words or word equivalents in the text portion of the message
B. The call sign of the originating station
C. A list of stations that have relayed the message
D. A box on the message form that indicates that the message was received and/or relayed

Careful handling of format traffic requires ham radio message handlers to always include a "check" to make sure that all the ***words in a message*** indeed were received in their entirety. **ANSWER A.**

Tracy, WM6T, checks the word count on an incoming message.

Website Resources

IF YOU'RE LOOKING FOR	THEN VISIT
All about ARES	www.ARRL.org/ARES
All about RACES	www.usraces.org
Emergency comms training	www.arrl.org/emergency-communications-training
ARRL emergency volunteers	www.ARRL.org/Volunteer
Military radio groups	www.netcom.Army.mil/MARS www.NAVYMARS.org
Amateur Radio audio clips from devastating 1964 Good Friday earthquake in Valdez, Alaska	https://youtu.be/rw9HZMagcb8
Website about radio amateurs during the 2004 tsunami in Indonesia	http://www.qsl.net/ab2qv/ares-tsunami.htm

Watch This! **https://www.youtube.com/playlist?list=PL1KAjn5rGhizk5f2_whBLJaOHj59OCI_3 (videos 243, 23, 133, 311, + 336)**

Weak Signal Propagation

T3C11 Why is the radio horizon for VHF and UHF signals more distant than the visual horizon?

A. Radio signals move somewhat faster than the speed of light
B. Radio waves are not blocked by dust particles
C. The atmosphere refracts radio waves slightly
D. Radio waves are blocked by dust particles

Now if you and your pal, walking on water, lost sight of each other's smiling face at 5 miles because of the curvature of the Earth, you could probably keep yakking on your 2-meter handhelds an ***additional couple of miles thanks to the refractive index of air on radio signals***, which makes the Earth seem less curved. So if you do your line-of-sight calculations, add another 15 percent to your visual range because your radio signals will be heard loud and clear beyond the visual horizon. This additional "mileage" is applied to both of your signals, so the actual "beyond horizon benefit" is about 30%. This is the origin of the common "4/3 Earth" principle for determining VHF distance. **ANSWER C.**

T3A12 What is the effect of fog and rain on signals in the 10 meter and 6 meter bands?

A. Absorption
B. There is little effect
C. Deflection
D. Range increase

Fog and rain have very little effect on most radio signals below UHF. 220 MHz is one possible exception, where precipitation just begins to have a noticeable effect. Fog and rain, on the other hand, can have tremendous effects (usually bad ones) on microwave frequencies. **ANSWER B.**

T3A07 What weather condition might decrease range at microwave frequencies?

A. High winds
B. Low barometric pressure
C. Precipitation
D. Colder temperatures

At microwave frequencies, the ***water molecules in the air*** can absorb significant amounts of RF energy. This is especially true at about 2.5 GHz (the frequency of microwave ovens) where water molecules are actually resonant. But absorptive events can occur well outside that frequency range. **ANSWER C.**

T3A02 What is the effect of vegetation on UHF and microwave signals?

A. Knife-edge diffraction
B. Absorption
C. Amplification
D. Polarization rotation

When a radio signal encounters a lossy substance, such as foliage, ***the signal energy is absorbed and turned into heat***. This absorption causes signal attenuation, and generally increases with frequency, but can be extreme when the radio frequency approaches that of water molecules, about 2.4 GHZ, the frequency of a microwave oven. The attenuation can be almost complete near this frequency, with just a small amount of water content in the leaves or bark. **ANSWER B.**

T3C01 Why are simplex UHF signals rarely heard beyond their radio horizon?

A. They are too weak to go very far
B. FCC regulations prohibit them from going more than 50 miles
C. UHF signals are usually not propagated by the ionosphere
D. UHF signals are absorbed by the ionospheric D region

The 6- and 10-meter bands are an exception to Technician Class privileges on frequency bands that usually propagate signals that travel line of sight. On the 6- and 10-meter bands, during the summer months, signals will many times refract off the E-layer of the ionosphere, and come back down hundreds and thousands of miles away. However, on 2 meters and the higher ***UHF*** frequencies, ***signals are usually not refracted by the ionosphere*** and head right off into outer space. **ANSWER C.**

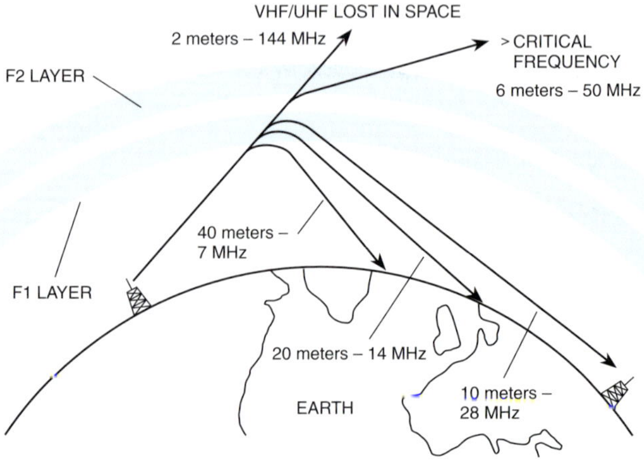

Critical Frequency
Source: *Antennas — Selection and Installation,* © 1986, Master Publishing, Inc.

T3C05 Which of the following effects may allow radio signals to travel beyond obstructions between the transmitting and receiving stations?

A. Knife-edge diffraction
B. Faraday rotation
C. Quantum tunneling
D. Doppler shift

You and a pal just earned your Technician Class licenses and you each have your new handhelds tuned to a simplex frequency. Your friend lives on the other side of a jagged rocky hill. Will you be able to communicate by simplex (no repeater) over this hill? Likely, yes. VHF FM signals from base stations, handhelds, and mobile radios normally propagate vertical polarization. When the lower edge of the vertical wave strikes a distant, elevated sharp object, like the peak between you and your friend, the lower portion of the wave tends to drag over the hill causing the upper portion of the wave front to bend right down to where your friend is waiting to hear from you. ***Knife-edge diffraction*** caused by an intervening building or hill may give you some great communications fun with your new Technician Class amateur radio license! **ANSWER A.**

Knife-Edge Diffraction

T3C06 What type of propagation is responsible for allowing over-the-horizon VHF and UHF communications to ranges of approximately 300 miles on a regular basis?

A. Tropospheric ducting
B. D region refraction
C. F2 region refraction
D. Faraday rotation

It is very common to be able to communicate through a repeater several hundred miles away during summer months. With a stationary high pressure system sitting over a large section of the country, atmospheric "layering" may trap warm air forming ***tropospheric "ducts" that refract VHF and UHF radio signals*** well beyond line of sight. This warm air inversion is what gives us longer than usual contacts, even with a handheld radio! **ANSWER A.**

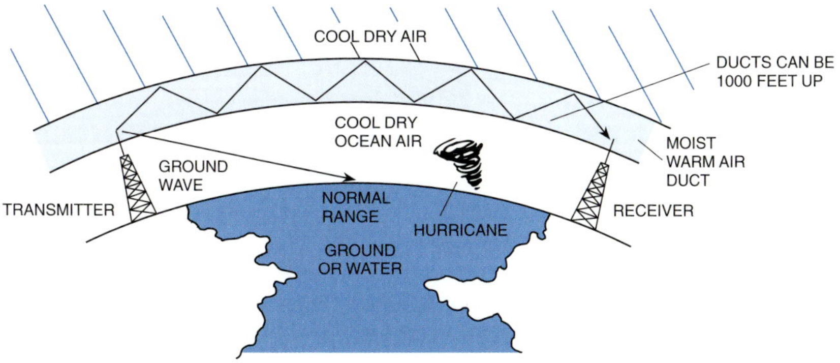

Tropospheric Ducting

T3C08 What causes tropospheric ducting?

A. Discharges of lightning during electrical storms
B. Sunspots and solar flares
C. Updrafts from hurricanes and tornadoes
D. Temperature inversions in the atmosphere

Ever see a mirage? This is frequently seen as water shimmering on a roadway, but it is actually the blue sky that you see on an inferior mirage. Another type of mirage may occur above us, called a superior mirage. This allows you to see things suspended upside down, many miles away! The superior mirage also creates longer than usual range on the 2 meter and 70 cm bands, thanks to a ***temperature inversion*** when a layer of warm air traps colder air below it, creating a tropospheric duct. Unlike tropospheric scattering, tropospheric ducting can be a very low-loss mode, allowing tremendous long-distance coverage with minimal power. (My first and shocking experience with this was working a Hawaiian station from the coast of California with a 2 meter handheld. Alas, we never see this kind of propagation in Alaska!) **ANSWER D.**

T3C03 What is a characteristic of VHF signals received via auroral backscatter?

A. They are often received from 10,000 miles or more
B. They are distorted and signal strength varies considerably
C. They occur only during winter nighttime hours
D. They are generally strongest when your antenna is aimed west

VHF signals on the 2 meter band travel "line of sight." The higher up you are from the surface of the Earth, the better your range. An interesting phenomenon, called an auroral backscatter reflection, may influence 2 meter single sideband and FM signals. Six meter and 2 meter radio waves will bounce off an auroral curtain as if they were bounced off a steel building! The ***incoming signals from a distant station will sound fluttery and distorted***. This is the unmistakable sound of an auroral bounce that can be heard hundreds of miles away! This characteristic flutter is not limited to VHF but is common on HF as well. Listen to an auroral radio call on the "On the Air" audio course you downloaded from **www.w5yi.org.** **ANSWER B.**

The auroral curtain will reflect radio signals over hundreds of miles, and when received they sound like a rapid flutter. Fascinating!

T3C07 What band is best suited for communicating via meteor scatter?

A. 33 centimeters
B. 6 meters
C. 2 meters
D. 70 centimeters

The Leonids and Geminids meteor showers are fun for hams! As a new Technician Class operator you can work the entire ***6 meter band***, running up to 1500 watts of power output. This is plenty of power to bounce a signal off a meteor trail or even the Moon! Hear the sound of this on the "On the Air" audio course downloaded from **www.w5yi.org.** **ANSWER B.**

The Effect of the Ionosphere on Radio Waves

To help you with the questions on radio wave propagation, here is a brief explanation of the effect the ionosphere has on radio waves.

The ionosphere is the electrified atmosphere from 40 miles to 400 miles above the Earth. You can sometimes see it as "northern lights." It is charged up daily by the sun, and does some miraculous things to radio waves that strike it. Some radio waves are absorbed during daylight hours by the ionosphere's D layer. Others are bounced back to Earth. Yet others penetrate the ionosphere and never come back again. The wavelength of the radio waves determines whether the waves will be absorbed, refracted, or will penetrate and pass through into outer space. Here's a quick way to memorize what the different layers do during day and nighttime hours:

The D layer is about 40 miles up. The D layer is a Daylight layer; it almost disappears at night. D for Daylight. The D layer absorbs radio waves between 1 MHz to 7 MHz. These are long wavelengths. All others pass through.

The E layer is also a daylight layer, and it is very Eccentric. E for Eccentric. Patches of E layer ionization may cause some surprising reflections of signals on both high frequency as well as very-high frequency. The E layer height is usually 70 miles.

The F1 layer is one of the layers farthest away. The F layer gives us those Far away signals. F for Far away. The F1 layer is present during daylight hours, and is up around 150 miles. The F2 layer is also present during daylight hours, and it gives us the Furthest range. The F2 layer is 250 miles high, and it's the best for the Farthest range on medium and short waves. At nighttime, the F1 and F2 layers combine to become just the F layer at 180 miles. This F layer at nighttime will usually bend radio waves between 1 MHz and 15 MHz back to earth. At night, the D and E layers disappear.

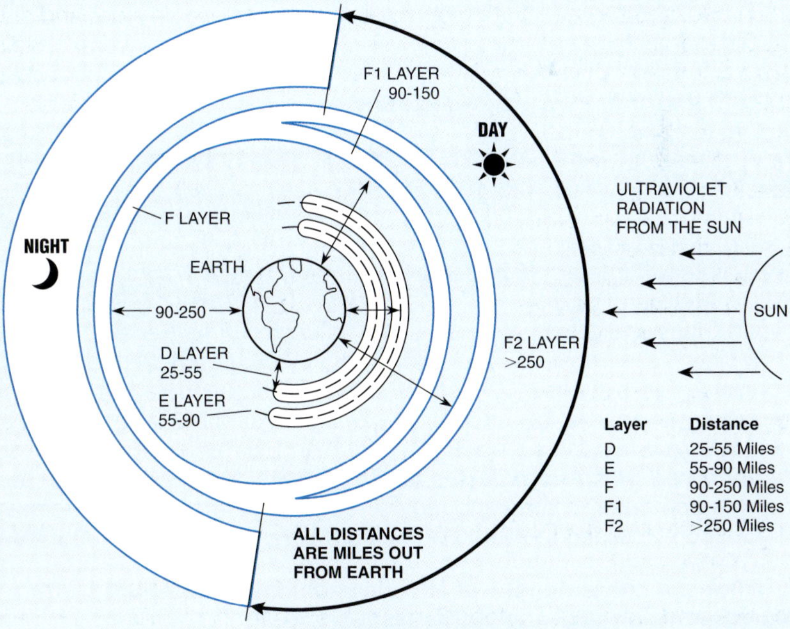

Ionosphere Layers

Source: *Antennas — Selection and Installation,* © 1986, Master Publishing, Inc., Niles, Illinois

5| 657-7300 8130 - 3:18

T3A11 Which region of the atmosphere can refract or bend HF and VHF radio waves?

A. The stratosphere
B. The troposphere
C. The ionosphere
D. The mesosphere

It is the ionosphere that gives us propagation all over North America on the 6- and 10-meter bands. We are steadily climbing up powerful Solar Cycle 25, with a couple more years to reach the peak! Enjoy sporadic E skip on these two bands every summer thanks to ***the ionosphere!*** **ANSWER C.**

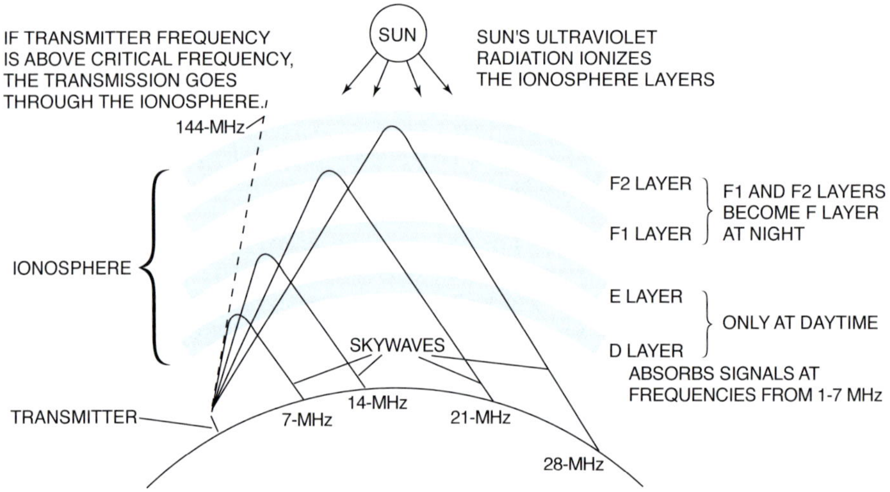

Ionosphere and Its Layers

T1B06 On which HF bands does a Technician class operator have phone privileges?

A. None
B. 10 meter band only
C. 80 meter, 40 meter, 15 meter, and 10 meter bands
D. 30 meter band only

Technician Class operators have some very valuable HF privileges. The entire former Novice Class frequency allocations have been grandfathered into the Technician Class license. 10 meter phone is a tremendously fun band to operate during any part of the sunspot cycle, and is capable of easy worldwide communications when the sunspots are "hot." And, as far as HF goes, while you are ***only allowed phone privileges on 10 meters***, there is nothing to stop you from using a computer to send and receive CW on 80, 40, and 15 meters as well, even if you aren't inclined to learn CW the hard way. [97.301(e), 97.305] **ANSWER B.**

T3C02 What is a characteristic of HF communication compared with communications on VHF and higher frequencies?

A. HF antennas are generally smaller
B. HF accommodates wider bandwidth signals
C. Long-distance ionospheric propagation is far more common on HF
D. There is less atmospheric interference (static) on HF

The "classical" or "standard" means of ***long-distance radio propagation is by means of HF (3-30 MHz, or shortwave) radio and ionospheric refraction***. While exceptional, long-distance propagation is possible on VHF (30-300 MHz), it is called "exceptional" for a reason. For the most part, VHF/UHF propagations is line of sight, and is quite similar to optical propagation. Weak signal work at VHF and above usually uses modes other than FM, such as CW, SSB, or digital modes. **ANSWER C.**

T1B10 Where may SSB phone be used in amateur bands above 50 MHz?

A. Only in sub-bands allocated to General class or higher licensees
B. Only on repeaters
C. In at least some segment of all these bands
D. On any band if the power is limited to 25 watts

As a new Technician, you will want to explore all that's available to you, and weak signal SSB (or CW!) operation is ***permitted in at least some portion of all the amateur bands above 50 MHz***. Most current high frequency transceivers offer full 6 meter capabilities at 100 watts output! Some will also offer 2 meter and 70 cm SSB capabilities, too. [97.305(c)] **ANSWER C.**

T1A06 What is the FCC Part 97 definition of a beacon?

A. A government transmitter marking the amateur radio band edges
B. A bulletin sent by the FCC to announce a national emergency
C. A continuous transmission of weather information authorized in the amateur bands by the National Weather Service
D. An amateur station transmitting communications for the purposes of observing propagation or related experimental activities

During times of low sunspot activity, it's sometimes difficult to tell if a lack of signals is due to no propagation or the possibility that there is just nobody on the air. This is especially true of the upper HF frequencies, such as 15 or 10 meters, commonly known as the "high bands." Propagation paths can be very intermittent on these bands, and you can easily miss such sporadic openings. A large number of continuously operating ***beacon stations reside on most of the high frequency bands and are extremely valuable for studying propagation*** and identifying band openings. You can usually find a large cluster of beacons right around 28.200 to 28.300 MHz. A list of the oft-heard 10-meter beacons is found here: **www.qsl.net/wj5o/bcn.htm.** Beacon stations are most frequently CW stations, identifying by their call sign followed by /b. [97.3(a)(9)] **ANSWER D.**

T3C09 What is generally the best time for long-distance 10 meter band propagation via the F region?

A. From dawn to shortly after sunset during periods of high sunspot activity
B. From shortly after sunset to dawn during periods of high sunspot activity
C. From dawn to shortly after sunset during periods of low sunspot activity
D. From shortly after sunset to dawn during periods of low sunspot activity

Remember your 10 meter SSB privileges are from 28.300 to 28.500 MHz. Start out around 28.400 and give your best spirited CQ call. Likely, ***during the day, during times of high Sunspot activity***, you may get a response from another station halfway across the country! **ANSWER A.**

T3A09 Which of the following results from the fact that signals propagated by the ionosphere are elliptically polarized?

A. Digital modes are unusable
B. Either vertically or horizontally polarized antennas may be used for transmission or reception
C. FM voice is unusable
D. Both the transmitting and receiving antennas must be of the same polarization

Elliptically (or circularly) polarized radio signals can be received with either a vertically or horizontally polarized antenna. However, you can obtain significant improvement in reception by using a circularly polarized antenna that is properly matched to the sense of the incoming signal; that is, using a right-hand circularly polarized antenna to receive a right-hand circularly polarized signal. Conversely, if you use a circularly polarized antenna of the wrong sense, the signal can often be totally rejected. This phenomenon lends itself to a lot of exciting experimentation, since very few hams currently take advantage of circularly polarized antennas on HF. These two "senses" are officially known in the scientific community as the X and the O modes, for extraordinary and ordinary rays, respectively. (Nobody said physicists could spell!) **ANSWER B.**

T3C04 Which of the following types of propagation is most commonly associated with occasional strong signals on the 10, 6, and 2 meter bands from beyond the radio horizon?

A. Backscatter
B. Sporadic E
C. D region absorption
D. Gray-line propagation

On 10 meters we may communicate regularly via skywaves with stations 1,000 miles away, thanks to ***sporadic E skip***. On 6 meters, this phenomenon usually occurs several times a week during the summer and in December. On 2 meters, skip conditions may prevail once or twice a year during the summer. **ANSWER B.**

T3C10 Which of the following bands may provide long-distance communications via the ionosphere's F region during the peak of the sunspot cycle?

A. 6 and 10 meters
B. 23 centimeters
C. 70 centimeters and 1.25 meters
D. All these choices are correct

Easy, super-long-distance HF propagation is possible at the peak of the sunspot cycle when the F2 ionospheric layer is thick and stable. Just a few watts can gain you world-wide communications on 6 and 10 meters during these times. As we climb up Solar Cycle 25, solar activity will begin to increase, and the higher frequencies will become progressively more stable. While we're waiting for this to happen, all is not lost as there are other, intriguing things happening besides the "normal" F2 propagation. Sporadic E can provide low-loss communications on ***6 and 10 meters*** at any time ***during the solar cycle***. However, they don't call it "sporadic" for nothing! It's important to be listening to catch these rather fleeting episodes. **ANSWER A.**

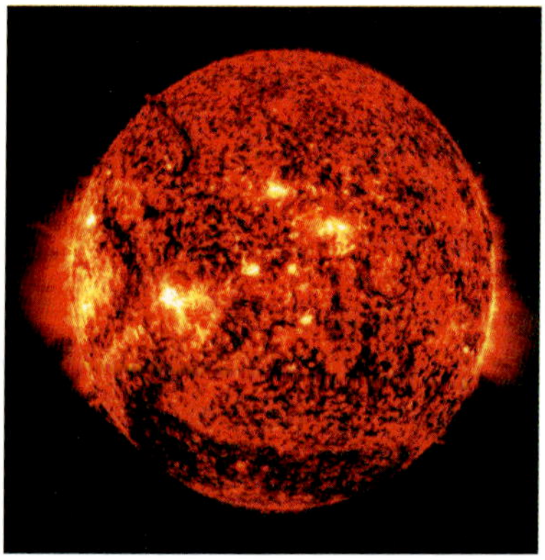

Solar flares and sunspots affect radio wave propagation.
Photo courtesy of N.A.S.A

T1B01 Which of the following frequency ranges are available for phone operation by Technician licensees?

A. 28.050 MHz to 28.150 MHz
B. 28.100 MHz to 28.300 MHz
C. 28.300 MHz to 28.500 MHz
D. 28.500 MHz to 28.600 MHz

Ten meters is the "on ramp" to long-distance, high frequency (HF) communications for most Technician Class licensees. New "techs" have quite a bit of "elbow room" on ten meters, a full two hundred kilohertz ***between 28.300 MHz and 28.500 MHz*** for voice. You also have a nice chunk of real estate, a whole *300 KHz,* for CW and data, between 28.000 and 28.300 MHz. As we are ascending a new sunspot cycle (Cycle 25) there will be plenty of good DX available for Technicians on 10 meters in the next few years. Listen for the many beacon stations around 28.300 to learn when radio conditions are good. You are encouraged to take advantage of our brand-new solar cycle! [97.301 (e)] **ANSWER C.**

T3B02 What property of a radio wave defines its polarization?

A. The orientation of the electric field
B. The orientation of the magnetic field
C. The ratio of the energy in the magnetic field to the energy in the electric field
D. The ratio of the velocity to the wavelength

Although the magnetic and electric fields of a radio wave are always at right angles to one another, by convention ***the polarization of a radio signal is defined by its electric field***. It is easy to remember this because the E field is the same orientation as the antenna ELEMENT that creates it. Horizontal wires have horizontal electric fields and require horizontally polarized antennas. Vertical wires have vertical electric fields and require vertically polarized antennas. **ANSWER A.**

T3A08 What is a likely cause of irregular fading of signals propagated by the ionosphere?

A. Frequency shift due to Faraday rotation
B. Interference from thunderstorms
C. Intermodulation distortion
D. Random combining of signals arriving via different paths

With exciting voice privileges on both 6 meters and 10 meters via ionospheric skip, Technician Class operators can work the World! Propagation on HF is different from what you will experience on a VHF handheld radio. Most long-distance radio paths are through the ionosphere and can take long, circuitous roundabout routes from Point A to Point B, sometimes several paths at the same time! As these signals refract via skywaves, they will constantly build up and then recede, like ocean waves coming in to shore. (As fate would have it, the other operator says his name and location just as the signal takes a brief nose-dive!) This constant coming and going of the skywave signal is due to selective fading where the ***incoming signal may arrive from several paths, combining in phase*** to be nice and strong, yet taking a quick deep fade as the signal comes in from different paths, out of phase. The good news with all this building and declining of signal strength from overseas skywave stations is that the fade outs are short in duration, leading to nice longer 10 seconds of solid reception. This in and out is also a good sign to you that the band is open for long range (DX) skywave contacts! **ANSWER D.**

Website Resources

IF YOU'RE LOOKING FOR	THEN VISIT
Educational site about propagation	prop.hfradio.org
Latest propagation conditions info	wap.hfradio.org
More on propagation	http://archive.is/www.haarp.alaska.edu wwww.bigear.org www.vlf.it www.arrl.org/weak-signal-vhf-dx-meteor-scatter-eme-moonbounce
6 Meter info	www.smirk.org
See Aurora up close	www.spaceweather.com
WWV propagation beacon info	www.nist.gov/pml/div688
Info on microwave groups	www.MicrowaveUpdate.org
More on microwave	www.Ham-Radio.com
The large network of Digisonde ionospheric sounders is probably the most useful real-time resource for radio propagation study and prediction. There is most likely a Digisonde near you!	http://digisonde.com/stationlist.php
Tomas Hood, NW7US, is one of the foremost propagation gurus in the U.S., if not the world. His HF propagation program, Proplab Pro3, is the best prediction software available to radio amateurs, by a large margin.	http://nw7us.us/
For the radio science geek, there is nothing quite like HAARP. I use the HAARP ionosonde daily to tell me whether it's worth firing up the HF rig or not.	http://www.gi.alaska.edu/facilities/haarp

Watch This! https://www.youtube.com/playlist?list=PL1KAjn5rGhizk5f2_whBLJaOHj59OCI_3 (videos 36, 111, 195, 320, + 318)

Talk to Outer Space!

T1A07 What is the FCC Part 97 definition of a space station?

A. Any satellite orbiting Earth
B. A manned satellite orbiting Earth
C. An amateur station located more than 50 km above Earth's surface
D. An amateur station using amateur radio satellites for relay of signals

While the common understanding of "space station" usually brings up images of large spacecraft with astronauts on board, the amateur radio definition is somewhat different. A number of very low orbiting satellites, such as CubeSats, fulfill the requirement of being an amateur space station. A space station doesn't even have to be in orbit, such as a short-duration sub-orbital rocket launch containing a radio payload. If it's ***50 km or more above the earth's surface, it's a space station.*** [97.3(a)(41)] **ANSWER C.**

Ham Hint: *If you're wondering how you can easily tune into the International Space Station, that is easy with a little 2-meter FM handheld. The International Space Station downlink, FM, is 145.800. When it is passing over, you might hear an astronaut and you might hear packet stations relaying messages globally. That's right, your little handheld is plenty powerful enough to pick up the International Space Station on an overhead pass!*

The International Space Station has a big ham station on board.
Photo courtesy of N.A.S.A.

T1B02 Which amateurs may contact the International Space Station (ISS) on VHF bands?

A. Any amateur holding a General class or higher license
B. Any amateur holding a Technician class or higher license
C. Any amateur holding a General class or higher license who has applied for and received approval from NASA
D. Any amateur holding a Technician class or higher license who has applied for and received approval from NASA

What a thrill to talk on your small handheld to an astronaut! It happens a lot, and with the ***Technician Class license or higher***, you, too, can explore space with astronauts and cosmonauts. [97.301, 97.207(c)] **ANSWER B.**

T8B10 What is a LEO satellite?

A. A sun synchronous satellite
B. A highly elliptical orbit satellite
C. A satellite in low energy operation mode
D. A satellite in low earth orbit

Low earth orbit (LEO) satellites (for the most part) are short lived, very inexpensive satellites. They are commonly used to launch Amateur Radio and related experiments. The Cube Sat program is one popular program that relies on LEO technology. LEOs move across the sky fast, and are a bit tricky to track, but they give amateurs an easy access to space technology. There are hundreds of LEOs around the world, and the International Space Station, (ISS) is one of the most prominent ones. **ANSWER D.**

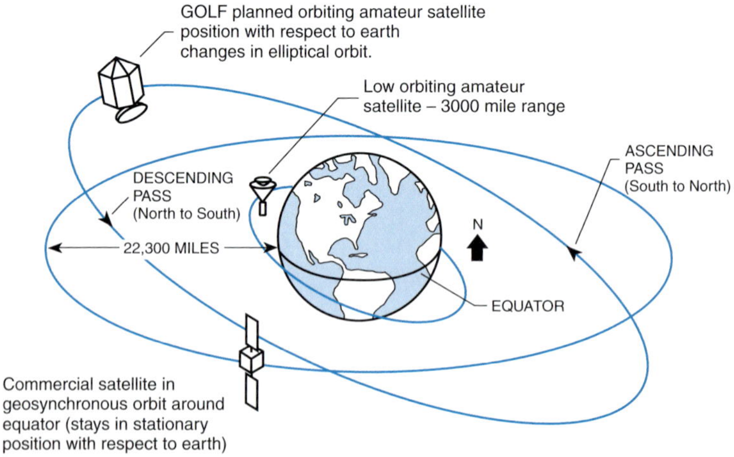

Ham Hint: *AMSAT reports that two of its GOLF (Greater Orbit Larger Footprint) series CubeSats are among 21 missions recommended for selection by NASA. These higher-altitude CubeSats will allow longer communications times. Get the latest news and learn about the dozens of small ham satellites in orbit right now from various countries, just waiting for you to operate on 2 meters/70cm bands with just a small hand held transceiver.* **www.amsat.org/**

Space Staion cross band repeater:
www.amsat.org/fm-satellite-frequency-summary/

T8B03 Which of the following are provided by satellite tracking programs?

A. Maps showing the real-time position of the satellite track over Earth

B. The time, azimuth, and elevation of the start, maximum altitude, and end of a pass

C. The apparent frequency of the satellite transmission, including effects of Doppler shift

D. All these choices are correct

"Keps" or ***Keplerian data sets*** are updated almost continually ***telling you everything about the satellite*** and allowing you to "find" any satellite there is – if you know how to interpret them. Modern satellite software takes all the drudgery out of this, making it almost "too easy." Varying levels of computer control are used by satellite enthusiasts, from merely informing the operator where to aim a handheld yagi antenna to fully controlling all the steering. Enjoy! **ANSWER D.**

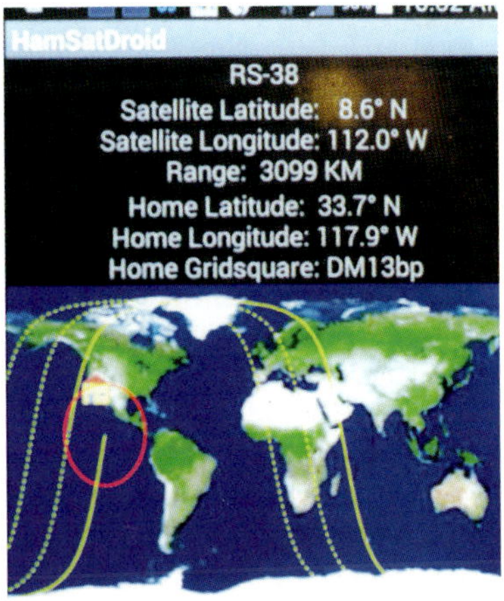

Computer programs and websites can show you where and when an amateur satellite or the Space Station will be in range of your ham station

---

T8B06 Which of the following are inputs to a satellite tracking program?

A. The satellite transmitted power

B. The Keplerian elements

C. The last observed time of zero Doppler shift

D. All these choices are correct

With the ***Keplerian elements*** properly inserted into your tracking program, the orbits of the satellites can be determined. These are sometimes known as "ephemeris" data, meaning they are continually changing unlike, say, the weight of the satellite. **ANSWER B.**

T8B05 What is a satellite beacon?

A. The primary transmit antenna on the satellite

B. An indicator light that shows where to point your antenna

C. A reflective surface on the satellite

D. A transmission from a satellite that contains status information

The satellite ***beacon is a continuous*** faint ***transmission that carries digitized information about the satellite*** itself. Ham satellite controllers can monitor the beacon with special software and check everything from the inside temperature to how fully-charged the satellite batteries are. **ANSWER D.**

T8B04 What mode of transmission is commonly used by amateur radio satellites?

A. SSB

B. FM

C. CW/data

D. All these choices are correct

The most popular and common mode of communications through an amateur radio satellite is FM, just like from your small handheld or mobile radio. On some satellites, we use SSB and CW/data from a base station multi-mode transceiver. These modes provide greater bandwidth for multiple conversations on the satellite's linear transponder. ***All of these answers*** will be fun! **ANSWER D.**

T8B11 Who may receive telemetry from a space station?

A. Anyone
B. A licensed radio amateur with a transmitter equipped for interrogating the satellite
C. A licensed radio amateur who has been certified by the protocol developer
D. A licensed radio amateur who has registered for an access code from AMSAT

It is pretty much a cardinal rule that once a radio signal has been transmitted it becomes public domain. There ***are no restrictions on who can receive radio signals***, though there may be some legal restrictions on what you do with that information. (This is especially true with things like cell phones and the like. But anyone can receive anything they like). **ANSWER A.**

T8B01 What telemetry information is typically transmitted by satellite beacons?

A. The signal strength of received signals
B. Time of day accurate to plus or minus 1/10 second
C. Health and status of the satellite
D. All these choices are correct

Telemetry signals for most amateur satellites are used for "housekeeping" purposes so its control operator can ***monitor the health and status*** of the "bird" and are not meant for general ham "consumption." The signals that are normally accessed by the ham user are sometimes referred to as the "payload." **ANSWER C.**

T8B09 What causes spin fading of satellite signals?

A. Circular polarized noise interference radiated from the sun
B. Rotation of the satellite and its antennas
C. Doppler shift of the received signal
D. Interfering signals within the satellite uplink band

As you listen to an FM or SSB satellite downlink signal it will rapidly rise and fall in signal strength. This is called ***"spin fading"*** and ***is caused by the satellite rotating in space*** in order to keep its solar panels from overheating. This rotation in space makes the signals fade in and out. Spin fading may be greatly reduced by the use of a circularly polarized receiving antenna. **ANSWER B.**

Clint, K6LCS, demonstrates live ham radio satellite activity to local clubs.

T8B07 What is Doppler shift in reference to satellite communications?

A. A change in the satellite orbit
B. A mode where the satellite receives signals on one band and transmits on another
C. An observed change in signal frequency caused by relative motion between the satellite and Earth station
D. A special digital communications mode for some satellites

Talk to Outer Space!

Ham satellite operators are constantly turning the dial during a conversation to make up for Doppler shift. As the satellite is headed towards you, the received signal may appear a few kHz high. As it passes overhead, the signal is right on frequency. And as it moves away from you the signal will be several kHz lower. It is very noticeable using single sideband and even on FM, especially on the 70cm uplink or downlink. Experienced satellite operators pre-store memory channels that will compensate for ***Doppler shift*** then change between those channels as the satellite passes. Amazing – talking through a satellite where you can actually hear ***frequency changes*** as the signal is coming at you, or moving away from you, at the speed of light! Listen to this sound on the "On the Air" audio course! **ANSWER C.**

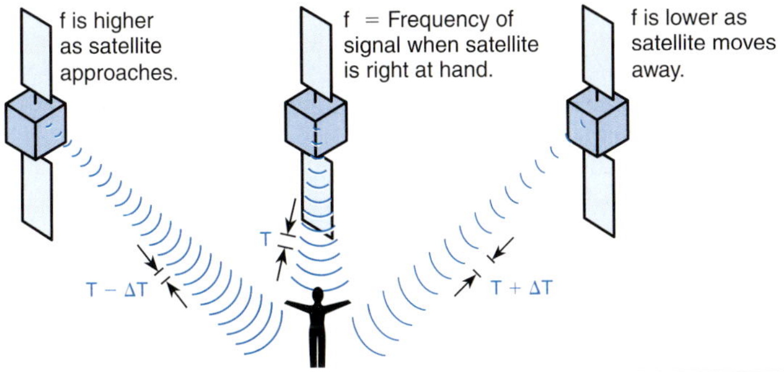

T8B12 Which of the following is a way to determine whether your satellite uplink power is neither too low nor too high?

A. Check your signal strength report in the telemetry data
B. Listen for distortion on your downlink signal
C. Your signal strength on the downlink should be about the same as the beacon
D. All these choices are correct

This is one of the primary purposes of ***satellite beacons,*** to give you a ***meaningful reference for adjusting your transmitter.*** Telemetry information is not generally available to the "consumer ham." ***Listening for your own signal is a good idea; it should be about the same strength as the beacon signal.*** (If you're hearing audible distortion on your retransmitted signal, you're "hitting" the satellite WAY too hard!) **ANSWER C.**

T8B02 What is the impact of using excessive effective radiated power on a satellite uplink?

A. Possibility of commanding the satellite to an improper mode
B. Blocking access by other users
C. Overloading the satellite batteries
D. Possibility of rebooting the satellite control computer

Some satellites, especially those working with SSB signals, use linear translators that produce an output power in direct proportion to the input power for each signal. There is only so much power to go around, and if one user is "hitting the bird" too hard their strong signal leaves less reserve power for everyone else, perhaps ***blocking access by other users***. It can also saturate the translator so that every other signal is highly distorted, or worse. Other satellites are essentially FM repeaters in the sky, which are somewhat more immune to this problem. However, the amateur radio prime directive of "use no more power than necessary" still applies. **ANSWER B.**

T8B08 What is meant by the statement that a satellite is operating in U/V mode?

A. The satellite uplink is in the 15 meter band and the downlink is in the 10 meter band
B. The satellite uplink is in the 70 centimeter band and the downlink is in the 2 meter band
C. The satellite operates using ultraviolet frequencies
D. The satellite frequencies are usually variable

An exciting Technician Class privilege is working ham radio satellites called OSCARs. You will need a full-duplex, dual-band radio with satellite tracking abilities to give you best access to those OSCARs that transmit and receive FM signals. If the mode is "U/V" it means we transmit on the satellite ***uplink frequency on the 70 cm "U" band***, and we listen to the ***satellite downlink on the 2 meter "V" band***. We hope all of you will support AMSAT, the nonprofit ham radio organization that sponsors our satellite programs. **ANSWER B.** Visit **www.AMSAT.org**

Satellite Sub-Bands	Frequency Range	Modes
High Frequency	21 - 30 MHz	Mode H
VHF	144 - 146 MHz	Mode V
UHF	435 - 438 MHz	Mode U
L band	1.26 - 1.27 GHz	Mode L
S band	2.4 - 2.45 GHz	Mode S
C band	5.8 GHz	Mode C
X band	10.4 GHz	Mode X
K band	24 GHz	Mode K

Website Resources

▼ IF YOU'RE LOOKING FOR	▼ THEN VISIT
Join AMSAT	www.amsat.org
Find Next Satellite Pass	www.Heavens-Above.com
Handheld Satellite Antenna	www.arrowantennas.com
Work satellites with a handheld	www.work-sat.com
Satellite tracking Software for Macs!.....	www.dogparksoftware.com
NASA Space Comms ISS	http://spaceflight.nasa.gov/station/ issfanclub.com/taxonomy/term/6 cubesat.org/index.php/documents/ developers
Space Station Contacts	www.ariss.org/current-status-of-iss-stations.html.
PSK 31 info DATA Readers	www.digipan.net

Watch This! **https://www.youtube.com/playlist?list=PL1KAjn5rGhizk5f2_whBLJaOHj59OCI_3 (videos 13 + 149)**

Going Digital

T8D09 What is CW?

A. A type of electromagnetic propagation
B. A digital mode used primarily on 2 meter FM
C. A technique for coil winding
D. Another name for a Morse code transmission

Next to spark transmission, ***CW (Morse Code) is the oldest form of radio communication*** and is still one of the most efficient and simple. Since the beginning of amateur radio CW has been the "default" mode and is allowed on every amateur band. It is implemented by simply turning a carrier wave signal on and off. **ANSWER D.**

Straight Key

T1B07 Which of the following VHF/UHF band segments are limited to CW only?

A. 50.0 MHz to 50.1 MHz and 144.0 MHz to 144.1 MHz
B. 219 MHz to 220 MHz and 420.0 MHz to 420.1 MHz
C. 902.0 MHz to 902.1 MHz
D. All these choices are correct

The lower frequency extremes or "bottom ends" of most amateur bands are restricted to CW only. As a Technician Class operator, you'll want to avoid the CW "windows" on the 6- and 2-meter bands ***(50.0 to 50.1 MHz and 144.0 to 144.1 MHz, respectively)*** if you're using a handheld radio. Better yet, learn some CW so you can use them! [97.305(a), (c)] **ANSWER A.**

T4A12 What is an electronic keyer?

A. A device for switching antennas from transmit to receive
B. A device for voice activated switching from receive to transmit
C. A device that assists in manual sending of Morse code
D. An interlock to prevent unauthorized use of a radio

With a conventional "straight" key, you must form every dot and dash manually, which normally limits your speed to about 25 words per minute. An electronic keyer takes a lot of the effort out of the process of ***manually sending Morse code*** by forming a string of perfectly timed dots and dashes, depending on which direction you move a paddle or pair of paddles. In addition, because keyer paddles use a side-to-side motion, rather than an up and down pumping motion, you have a lot less wear and tear on your wrist. There are a variety of paddles and associated electronics available, and it is a lot of fun to check out the vast assortment of these devices at places like the Dayton Hamvention. There are also semi-automatic mechanical keyers, such as the ancient-but-much-alive Vibroplex, which are still extremely popular amongst dedicated CW operators. **ANSWER C.**

T4A06 What signals are used in a computer-radio interface for digital mode operation?

A. Receive and transmit mode, status, and location
B. Antenna and RF power
C. Receive audio, transmit audio, and transmitter keying
D. NMEA GPS location and DC power

There are only three connections necessary between a transceiver and computer to create a first-rate digital communications station. You'll need one ***audio cable*** to go between the sound card output and the transceiver's microphone input (or preferably a rear panel line-in connector). You'll need ***another audio*** cable between the transceiver's speaker or headphone output (also, possibly a real panel audio connector) and the sound card microphone input. Finally, you'll need a DC cable between your computer's serial port and the transceiver's ***push-***

A typical modern ham shack. The "main" rig is a Ten Tec Jupiter, a software defined radio, and associated computer. A large antenna tuner and legal limit linear amplifier are at the right. And, most important, a large 24 hour clock – a vital accessory for any ham shack!

to-talk (PTT) connector, which is usually a pin on the microphone connector or sometimes a dedicated DIN connector on the rear panel. However, if your transceiver has a VOX (voice operated transmit) function, you can do away with the PTT connection entirely, by allowing the transmit audio itself to key the transmitter at the right time. **ANSWER C.**

T4A10 What function is performed with a transceiver and a digital mode hot spot?

A. Communication using digital voice or data systems via the internet
B. FT8 digital communications via AFSK
C. RTTY encoding and decoding without a computer
D. High-speed digital communications for meteor scatter

Wi-Fi routers and hot spots have made access to the internet simple and neat, without the need of cumbersome ethernet cables and such. ***An amateur digital hot spot is simply a very simple Wi-Fi router designed just for digital voice connections.*** A number of amateur internet protocols allow connection to repeaters of various types through the internet. **ANSWER A.**

T8D01 Which of the following is a digital communications mode?

A. Packet radio
B. IEEE 802.11
C. FT8
D. All these choices are correct

The oldest "official" digital mode is ***RTTY***, or CW with a small adjustment of the definition. So, too, are ***IEEE 802.11 and JT65***. What makes ***all of these*** modes digital is that the symbols are formed by discrete values of some parameter such as amplitude, frequency, or phase, as opposed to audio or video, which can assume any value. It looks like there is no end to the variety of new digital modes available for amateur experimentation. And inevitably, the latest and greatest digital mode will eventually be familiar and comfortable for the majority of hams such as the WSJT-X software suite, ***mode FT8***. **ANSWER D.**

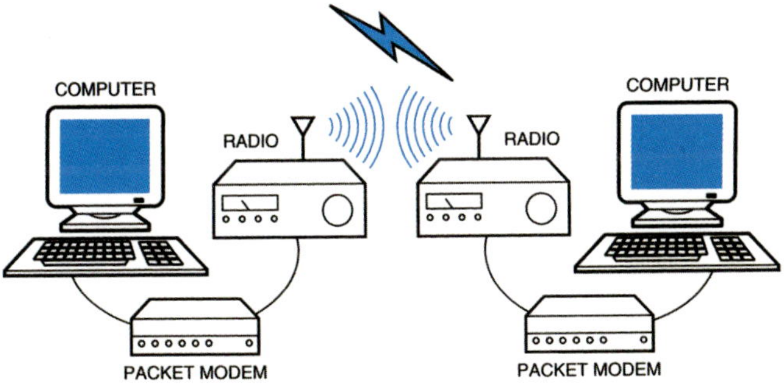

A Packet Radio System

T4A07 Which of the following connections is made between a computer and a transceiver to use computer software when operating digital modes?

A. Computer "line out" to transceiver push-to-talk
B. Computer "line in" to transceiver push-to-talk
C. Computer "line in" to transceiver speaker connector
D. Computer "line out" to transceiver speaker connector

This is a little bit of a trick question because you actually need TWO audio connections. However only one of these answer choices is correct. You can MONITOR digital modes with just the ***audio line from your rig's speaker output to the computer line input.*** **ANSWER C.**

T8D13 What is FT8?

A. A wideband FM voice mode
B. A digital mode capable of low signal-to-noise operation
C. An eight channel multiplex mode for FM repeaters
D. A digital slow-scan TV mode with forward error correction and automatic color compensation

FT8 and similar modes are not particularly suitable for casual conversations (QSO's), but have a very regimented, specific, and automatic format for exchanging signal reports and other relevant information. Think of FT8 as two radios talking to each other with actual human operators being somewhat extraneous. If you want to talk about the weather, sports, or homework, FT8 is probably not going to be too exciting. But if it's strictly making a contact under weak signal conditions, it's ***low signal-to-noise operation*** is hard to beat. You can get WAS (Worked All States) or DXCC (DX Century Club) in short order with FT8, even under the worst of conditions. **ANSWER B.**

T4A04 How are the transceiver audio input and output connected in a station configured to operate using FT8?

A. To a computer running a terminal program and connected to a terminal node controller unit
B. To the audio input and output of a computer running WSJT-X software
C. To an FT8 conversion unit, a keyboard, and a computer monitor
D. To a computer connected to the FT8converter.com website

Operating FT8 (or any other sound card mode) is extremely simple. All you need is WSJT-X software and a sound card. Most laptop sound cards are more than adequate for this. ***Just take the line level audio output of the transceiver and plug it into the MIC input of the sound card (or line input if it has one). Take the audio line INPUT of your transceiver and plug it into the sound card output (such as a headphone jack).*** It's best to keep the audio levels in both direction a bit on the low side. WSJT-X and other digital software has a TUNE provision for setting up the transmit levels. **ANSWER B.**

T8D10 Which of the following operating activities is supported by digital mode software in the WSJT-X software suite?

A. Earth-Moon-Earth
B. Weak signal propagation beacons
C. Meteor scatter
D. All these choices are correct

WSJT is a computer program used for weak-signal radio communication between amateur radio operators. All of the various JT modes were developed with weak signal work in mind, though they have achieved levels of popularity unexpected by the original developer. Working Moonbounce used to require immense antenna arrays and full legal limit power levels. With the new weak signal JT modes, a ham with a modest yagi antenna array and a hundred watts can work Moonbounce. It also can be used for meteor scatter, as well as to tune in to weak-signal propagation beacons. WSJT is a real "game changer" among ***all these modes***. The various "JT modes" are the invention of Nobel Prize winner, astrophysicist, and radio amateur of renown, Joseph Hooton Taylor Jr., W1JT. He is one of the many radio amateurs who have made major scientific discoveries over the past hundred years or so. What's stopping you from joining the ranks of these giants? **ANSWER D.**

T8D08 Which of the following is included in packet radio transmissions?

A. A check sum that permits error detection
B. A header that contains the call sign of the station to which the information is being sent
C. Automatic repeat request in case of error
D. All these choices are correct

Get your computers ready for some Technician Class fun. Packet radio is a great way to send blocks of information, quickly, over the airwaves. Your packet will contain a header with the call sign of the station you wish to reach, and several steps of error detection and error repeat requests. So, for this question, ***all of these*** features are included when you send packet radio. **ANSWER D.**

T8D11 What is an ARQ transmission system?

A. A special transmission format limited to video signals
B. A system used to encrypt command signals to an amateur radio satellite
C. An error correction method in which the receiving station detects errors and sends a request for retransmission
D. A method of compressing data using autonomous reiterative Q codes prior to final encoding

ARQ stands for Automatic Repeat Request, a system that lets your digital station automatically ask the other station to "say again" a missed set of characters! The earliest, commonly-used ARQ system in amateur radio was the largely defunct AMTOR system. The commercial version of AMTOR, SITOR, is still used on the high seas for third party radio traffic as well as weather bulletins. PACTOR has largely replaced AMTOR because it has a full ASCII character set, whereas AMTOR only had upper case letters. PACTOR and similar systems are capable of operating in the ARQ mode or the FEC (forward error correction) mode, depending on whether they are engaged primarily in two-way or one-way "broadcast" communications. **ANSWER C.**

T3A10 What effect does multi-path propagation have on data transmissions?

A. Transmission rates must be increased by a factor equal to the number of separate paths observed
B. Transmission rates must be decreased by a factor equal to the number of separate paths observed
C. No significant changes will occur if the signals are transmitted using FM
D. Error rates are likely to increase

You are going to have a blast sending data with your laptop and your ham set. When conditions are good, data exchanges are letter/number perfect! However, when ***multipath signals*** add and subtract to your signal, ***error rates are likely to increase***. Since the earliest days of data communications (which includes Morse code, by the way) hams have experimented with using multiple antennas and receivers to reduce multipath problems. This is still an area ripe for experimentation! **ANSWER D.**

T8D05 Which of the following is an application of APRS?

A. Providing real-time tactical digital communications in conjunction with a map showing the locations of stations
B. Showing automatically the number of packets transmitted via PACTOR during a specific time interval
C. Providing voice over internet connection between repeaters
D. Providing information on the number of stations signed into a repeater

Want to look at a ***GPS map and see where other ham stations are located*** around you? APRS can do that for you! APRS is an extension of "classic" amateur packet technology. It is an application-level protocol that rides on top of the lower "layers" of packet technology. Therefore, any packet TNC can be used to transport APRS data. **ANSWER A.**

Kenwood dual bander plugged into the Avmap G5 GPS position plotter.

T8D03 What kind of data can be transmitted by APRS?

A. GPS position data
B. Text messages
C. Weather data
D. All these choices are correct

APRS (Automatic Position Reporting System) was one of the first refinements of "plain vanilla" packet radio in the 1990s and is still immensely popular. When it first became available, very few hams had GPS access, so it relied on "dead reckoning" to supply the position data, which wasn't terribly accurate. Now with readily available, high precision GPS, APRS is better than ever. ***APRS can transmit GPS position data, text messages, and weather data.*** **ANSWER D.**

T8D06 What does the abbreviation "PSK" mean?

A. Pulse Shift Keying
B. Phase Shift Keying
C. Packet Short Keying
D. Phased Slide Keying

PSK stands for ***Phase Shift Keying***; and if you listen carefully, the little warble is indeed information passing over an ultra-narrow carrier. You can also hear the sounds of PSK on your HF radio by tuning to 14.070 MHz. **ANSWER B.**

Some high end transceivers have a built-in digital interface for PSK and RTTY operation. Nothing else to buy!

T8C11 What is an amateur radio station that connects other amateur stations to the internet?

A. A gateway
B. A repeater
C. A digipeater
D. A beacon

Hundreds of ham radio operators throughout the world have tied their home computers and ham sets into the Internet. The systems are called WinLink 2000, EchoLink, IRLP and DStar. These free ***gateway*** stations are fully automated to provide ham radio operators access to the Internet for sending and receiving e-mails. **ANSWER A.**

Ham Hint: *For the Technician Class operator, VHF and UHF gateway stations could allow the ham radio emergency communicator, in the field, to send messages via the Internet to an American Red Cross center or an EOC. This system also allows non-licensed hams to generate e-mail to you, going through a WinLink gateway station. And once you get your General Class license, we could cruise the world and always have e-mail capabilities over longer-range, high-frequency networks. While bandwidth limitations do not support surfing the web, you can download and send e-mails along with VHF/UHF medium-resolution diagrams and charts, too.*

T8C08 What is the Internet Radio Linking Project (IRLP)?

A. A technique to connect amateur radio systems, such as repeaters, via the internet using Voice Over Internet Protocol (VoIP)
B. A system for providing access to websites via amateur radio
C. A system for informing amateurs in real time of the frequency of active DX stations
D. A technique for measuring signal strength of an amateur transmitter via the internet

VoIP allows rather simple "gateway" methods between amateur radio and the commercial Internet. The ***IRLP uses Voice over Internet Protocol (VoIP) to connect repeaters***, allowing us to talk around the World on our little handheld radios. IRLP is fully open-sourced, as well, developed on the Linux operating system. Visit + **www.winsystem.org** to see over 80 repeaters in the U.S. and around the world tied together for IRLP excitement. **ANSWER A.**

T2B06 What type of signaling uses pairs of audio tones?

A. DTMF
B. CTCSS
C. GPRS
D. D-STAR

Dual Tone Multiple Frequency (DTMF), or Touch Tone®, is the system that is used on land-line phones (remember those?) to dial the numbers. It is also used for a number of signaling functions in amateur radio, such as performing remote control tasks, or even bringing up auto-patch repeaters from the touch pad on your hand-held. **ANSWER A.**

T8C06 How is over the air access to IRLP nodes accomplished?

A. By obtaining a password that is sent via voice to the node
B. By using DTMF signals
C. By entering the proper internet password
D. By using CTCSS tone codes

IRLP nodes are basically "phone numbers" which access the node in question. ***DTMF signals*** sent from the keypad of your handheld or mobile microphone are the most practical way of ***accessing these nodes*** across "boundaries" between commercial and amateur communications networks. Although this is not the only conceivable way of doing this, the IRLP standard calls for using standard DTMF tones. By the way, DTMF stands for Dual Tone Multi-Frequency, and are the same tones used by your touch-tone telephone. **ANSWER B.**

Almost all VHF/UHF handhelds have a key pad for entering IRLP node numbers. Many mobile VHF/UHF radios may take an optional keypad microphone to generate these dual tones on transmit.

T8C07 What is Voice Over Internet Protocol (VoIP)?

A. A set of rules specifying how to identify your station when linked over the internet to another station
B. A technique employed to "spot" DX stations via the internet
C. A technique for measuring the modulation quality of a transmitter using remote sites monitored via the internet
D. A method of delivering voice communications over the internet using digital techniques

One of the most amazing things about the TCP/IP protocol, around which the entire Internet was "wrapped," is how well it adapts to new technologies not even remotely thought of when it was first put together. ***Voice over IP uses well-established TCP/IP packet switching to send packets of voice*** over exactly the same hardware as everything else Internet related. **ANSWER D.**

T8C09 Which of the following protocols enables an amateur station to transmit through a repeater without using a radio to initiate the transmission?

A. IRLP
B. D-STAR
C. DMR
D. EchoLink

EchoLink allows hams to access hundreds of repeaters around the world with just a computer or a smart phone. You must first register with EchoLink with a legitimate call sign, which prevents unlicensed internet users from accessing the system. Once you're registered, you can search through countless repeaters around the world and talk to them just as if you had a handheld radio in the area. **ANSWER D.**

T8C10 What is required before using the EchoLink system?

A. Complete the required EchoLink training
B. Purchase a license to use the EchoLink software
C. Register your call sign and provide proof of license
D. All these choices are correct

When you ***register*** your call sign ***with the Echolink system***, you assume the role of a control operator, which is why the Echolink system must ***verify that you are an actual licensed ham***. Even if you don't have an actual transmitter within a hundred miles of your desk, when you become an Echolink control operator, you are responsible for what goes over that remote repeater just as if you were at the transmitter site. **ANSWER C.**

T8D04 What type of transmission is indicated by the term "NTSC?"

A. A Normal Transmission mode in Static Circuit
B. A special mode for satellite uplink
C. An analog fast-scan color TV signal
D. A frame compression scheme for TV signals

Here in the U.S., ***NTSC is a standard for fast-scan analog color television signals*** used by ham radio ATV stations. TV broadcasters have made the switch to all-digital, so hams are the last ones to stick with the tried-and-proven NTSC analog transmissions. So don't throw out your old analog TV set just yet – with a simple $99 down-converter, fast-scan ham television in full color awaits you! (FYI, NTSC stands for National Television System Committee.) **ANSWER C.**

Weekly nets may bring up dozens of hams on ATV, showing off their station, as well as the area of ATV coverage.

T8A10 What is the approximate bandwidth of AM fast-scan TV transmissions?

A. More than 10 MHz
B. About 6 MHz
C. About 3 MHz
D. About 1 MHz

A fascinating type of ham radio operation is called fast-scan television, abbreviated "ATV." Hams can send fast-scan TV pictures direct or through special fast-scan television repeaters over hundreds of miles! Years ago, the Pasadena Rose Parade was covered with fast-scan TV pictures keeping officials up-to-date on float safety. Fast-scan TV is very wide – ***about 6 MHz*** – and can be found on 70-centimeter frequencies and higher where equipment cost is modest, especially if you already own a video camera. See ATV action for yourself at + **www.atn-tv.org** and follow their gradual switch to high-definition digital TV transmission and reception for ham radio use. **ANSWER B.**

Website Resources

▼ IF YOU'RE LOOKING FOR	▼ THEN VISIT
FLDIGI is the most popular digital HF mode software, and is usually all you need to explore all the different digital sound card modes	http://www.w1hkj.com/
MULTIPSK is another all-digital-mode suite. It probably should win an award for the world's ugliest user interface, but it works GREAT on every mode, and has a few modes that FLDIGI doesn't have, such as SSTV. It has been my primary digital ham software since such things existed. It has a very full set of APRS tools, as well. For the Technician, I recommend MULTIPSK over FLDIGI, but that's just one ham's opinion	http://f6cte.free.fr/index_anglais.htm
All about amateur TV	www.atn-tv.org/
Microwave experimenters TV system	www.qsl.net/ki7cx/aatv_f.htm
70 cm ATV info	www.downeastmicrowave.com
Free subscription to ATV newsletter	atv-newsletter@hotmail.com
Using FT8	https://youtu.be/YyWX0i87P0o
CW Paddles	www.I2RF.com

Watch This! **https://www.youtube.com/playlist?list=PL1KAjn5rGhizk5f2_whBLJaOHj59OCI_3 (videos 298, 8, + 71)**

CD 3 TRACK 4

Multi-Mode Radio Excitement

T7A02 What is a transceiver?

A. A device that combines a receiver and transmitter
B. A device for matching feed line impedance to 50 ohms
C. A device for automatically sending and decoding Morse code
D. A device for converting receiver and transmitter frequencies to another band

For convenience, compactness, and economy, most modern amateur radio equipment comes in the form of a ***"transceiver,"*** which ***combines both transmitting and receiving functions*** in one package. However, many experienced radio amateurs prefer the performance and versatility of separate transmitters and receivers (also known as "separates" or "twins"), where the vastly different functions of transmitting and receiving can be thoroughly optimized. In addition, many radio amateurs enjoy using vintage or "boat anchor" radios, built when the concept of combining both transmitters and receivers into one cabinet was unheard of. In fact, many manufacturers during the Golden Age of ham radio built only transmitters or receivers, and the typical ham was compelled to put together his station from a variety of specialized sources or build his own. Count yourself lucky to be a radio amateur now! **ANSWER A.**

A multi-mode transceiver offers many modulation and operating modes in a single unit. These can include SSB, CW, VHF/UHF, and digital modes.

T8A03 Which type of voice mode is often used for long-distance (weak signal) contacts on the VHF and UHF bands?

A. FM
B. DRM
C. SSB
D. PM

Unlike the common FM mode that you will be using with your dual-band handheld, ***single sideband (SSB)*** requires special – yet not too expensive – equipment. 6-meter single-sideband radios are seen regularly at swap meets. 6-meter SSB signals are ideal for bouncing off the ionosphere. And, as you will hear on the "On the Air" audio course, it is mighty exciting! **ANSWER C.**

What's a multi-mode transceiver?

It's a transceiver that can send and receive different modes of radio signals. The term "mode" refers to the type of modulation emission used to send signals. Look at the illustration below. We send Morse code signals using CW, continuous wave. When you tap out a message on a code key, you interrupt (turn off and on) the unmodulated carrier continuous wave. In AM modulation, the RF carrier wave is modulated by the information signal and the amplitude (height) of the signal varies. SSB is single sideband, a form of AM, amplitude modulation. FM is frequency modulation, and that's what most VHF/UHF handheld radios use. In FM, the frequency of the carrier wave is changed by the information signal. To learn more about how radios work, we suggest **Basic Communications Electronics** *by Jack Hudson, W9MU and Jerry Luecke, KB5TZY, available from The W5YI Group, 800-669-9594 or* **www.w5yi.org.**

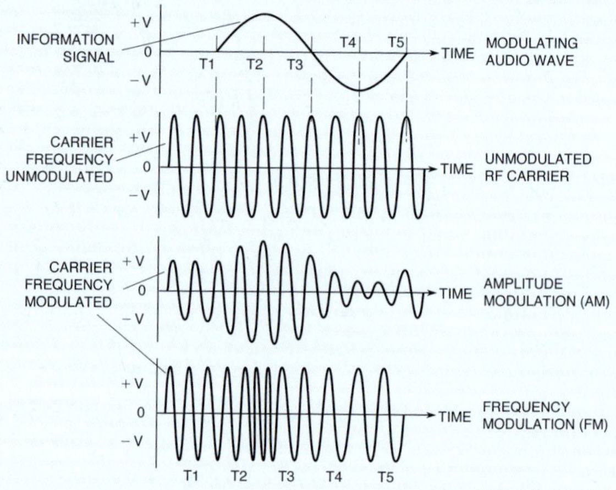

T8A07 What is a characteristic of single sideband (SSB) compared to FM?

A. SSB signals are easier to tune in correctly
B. SSB signals are less susceptible to interference
C. SSB signals have narrower bandwidth
D. All these choices are correct

We use frequency modulation through FM repeaters on many popular VHF and UHF ham bands. But near the bottom of 6 meters, 2 meters, 1 1/4 meters, and at 432 and 1296 MHz, we use ***single sideband***, which ***uses less bandwidth than FM signals***. **ANSWER C.**

T8A12 Which of the following is a disadvantage of FM compared with single sideband?

A. Voice quality is poorer
B. Only one signal can be received at a time
C. FM signals are harder to tune
D. All these choices are correct

FM has a property known as the "capture effect" that causes ***only the stronger of two signals on the same frequency to be detected***. While this makes FM less subject to interference, it makes weak signal work much more difficult. And it can also have disastrous effects. Aircraft radios all use AM (of which SSB is a type), and even though it is noisier, they will continue to do so. An aircraft controller (as well as multiple aircraft) absolutely MUST hear everyone else at all times. FM's capture effect would be absolutely deadly in the sky. **ANSWER B.**

T8A01 Which of the following is a form of amplitude modulation?

A. Spread spectrum
B. Packet radio
C. Single sideband
D. Phase shift keying (PSK)

Single sideband (SSB) is an efficient, modified form of ***amplitude modulation*** (AM), where the audio signal causes a change in RF output power. This is in contrast with FM, where the transmitter output power remains constant regardless of modulation (or lack thereof). SSB is much more suitable for weak signal work than FM and is the standard voice mode for long-distance HF communications. **ANSWER C.**

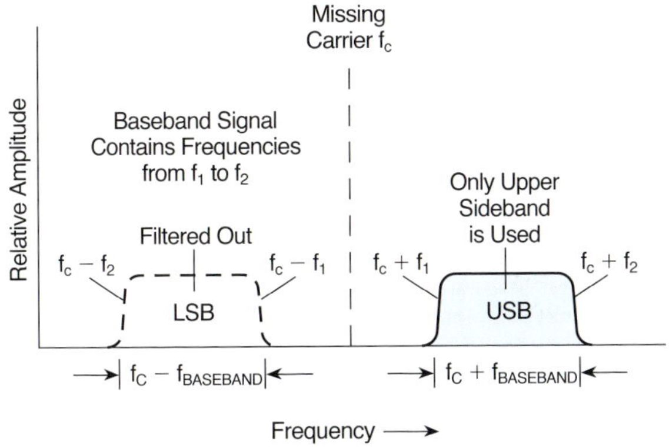

SSB signals are Amplitude Modulated (AM) with the carrier and one sideband suppressed.

T8A06 Which sideband is normally used for 10 meter HF, VHF, and UHF single-sideband communications?

A. Upper sideband
B. Lower sideband
C. Suppressed sideband
D. Inverted sideband

On VHF and UHF, by convention, ***we always use upper sideband*** as our communications mode. Single sideband is an advanced feature for you to try after you have been on the air for a few months as a Technician Class operator. You won't find single sideband capability in a ham handheld. You will find SSB in high frequency radios. This equipment may also give you 6 meters, 2 meters, and 70 cm, in addition to the worldwide HF capability. We don't use single sideband through repeaters. SSB is found at the bottom of most VHF/UHF ham bands, so you will want to try the following frequencies – all upper sideband – to hear some SSB signals:

10 meters	28.400 MHz SSB
6 meters	50.125 MHz SSB
2 meters	144.200 MHz SSB
70 cm	432.100 MHz SSB

No FM allowed on these weak-signal SSB frequencies. Equipment that includes all of these bands plus HF costs less than $1,000. **ANSWER A.**

The Icom HF/VHF/UHF transceiver running on a small external battery with Gordo holding a small volt and amp meter.

T7A09 What is the function of the SSB/CW-FM switch on a VHF power amplifier?

A. Change the mode of the transmitted signal
B. Set the amplifier for proper operation in the selected mode
C. Change the frequency range of the amplifier to operate in the proper segment of the band
D. Reduce the received signal noise

Any form of AM (amplitude modulated) signal, such as full carrier AM, SSB, or analog video, must be amplified with a linear amplifier. FM and pure frequency-shifted modes such as RTTY can be amplified with a highly efficient Class C amplifier. Most modern ***VHF external power amplifiers can be switched between linear and Class C for the appropriate mode***. CW is an interesting case in this regard. You can amplify CW with a Class C amp, but it will "harden" the keying waveshape, sometimes enough to cause severe key clicks. To "properly" amplify a CW signal without altering the waveshape, you should use a linear amplifier. **ANSWER B.**

T7A08 Which of the following describes combining speech with an RF carrier signal?

A. Impedance matching
B. Oscillation
C. Modulation
D. Low-pass filtering

Modulation is the process of applying information (voice, video, or digital) onto a radio signal. The information modifies some characteristic of the radio frequency carrier so that the information can be transmitted through the air. Inside that brand new dual-band handheld is a circuit called the modulator. It converts your spoken word going into the microphone into an electrical speech signal that is combined with the RF carrier. The modulator is always on the transmit side of your equipment, tied in to the microphone. **ANSWER C.**

T8A08 What is the approximate bandwidth of a typical single sideband (SSB) voice signal?

A. 1 kHz
B. 3 kHz
C. 6 kHz
D. 15 kHz

The properly adjusted, ***single sideband*** transmitter on VHF and UHF frequencies for Technician Class operators occupies approximately ***3 kHz of bandwidth***, depending on your voice characteristics. By convention, we employ upper sideband for VHF and UHF weak-signal work. Be sure to listen to the "On the Air" audio course downloaded from **www.w5yi.org** for the sounds of SSB. **ANSWER B.**

SSB and CW filters inside a transceiver.

Ham Hint: *Handhelds do not incorporate a single-sideband capability, but some "multimode" mobiles do include SSB operation. Listen to the "On the Air" audio course and discover the fascinating world of longer range transmissions over hundreds and sometimes thousands of miles using single sideband. Listen to the regular summertime occurrences of long-range skywave SSB signals on the 6- and 10-meter bands! After you listen, hopefully you will get very excited about all the fun that single sideband can add to your ham radio activity beyond your daily communications on regular FM repeaters.*

T4B06 Which of the following controls could be used if the voice pitch of a single-sideband signal returning to your CQ call seems too high or low?

A. The AGC or limiter
B. The bandwidth selection
C. The tone squelch
D. The RIT or Clarifier

If you plan to do some weak signal work, great! We need more weak signal operators using CW (Morse code) and single sideband (SSB). It will take a little practice to tune in voice properly on single sideband. You should set your ***receiver RIT or clarifier*** to the neutral position, and then slightly adjust the big tuning knob for proper SSB voice reception. You can then make fine tuning adjustments with the receiver RIT or clarifier knob to raise or lower a signal pitch that is wandering a bit from an older, unstable radio. **ANSWER D.**

T7A05 What is the name of a circuit that generates a signal at a specific frequency?

A. Reactance modulator
B. Phase modulator
C. Low-pass filter
D. Oscillator

All radio signals originate with an ***oscillator*** of some sort, somewhere in the circuitry. An oscillator is essentially an amplifier with positive feedback and some frequency-determining components, most frequently a resonant or tuned circuit. Oscillators come in countless varieties, using everything from vibrating quartz crystals to wobbling atoms as the resonant circuit. Some of the more classic oscillator circuits used in amateur radio are the Colpitts, the Hartley, and the Pierce oscillators. Various "multivibrator" circuits also function in an oscillator-like fashion in many digital circuits. **ANSWER D.**

The insides of your handheld and mobile radio equipment feature a sturdy circuit board, where robotics assemble and test components as they are wave soldered on the board. With components this closely spaced, keep your radio gear bone dry!

T8A05 Which of the following types of signal has the narrowest bandwidth?

A. FM voice
B. SSB voice
C. CW
D. Slow-scan TV

FM voice and slow-scan TV use modest bandwidth; SSB voice has a "skinny" bandwidth, and the narrowest bandwidth of all these answer choices is ***Morse code, CW***. Even though the code test for CW has been eliminated for all classes of ham licenses, CW will always be a popular, ultra-narrow-bandwidth way of communicating. When all else fails, CW gets through! **ANSWER C.**

T8A11 What is the approximate bandwidth required to transmit a CW signal?

A. 2.4 kHz
B. 150 Hz
C. 1000 Hz
D. 15 kHz

We hope you will try some CW on the worldwide portion of your Technician Class privileges on 80, 40, 15 and 10 meters. When you switch your multi-mode rig over to CW you will see that the bandwidth narrows. This minimizes interference from other CW signals off frequency. On transmit, ***150 Hz*** is just right. **ANSWER B.**

T4B10 Which of the following receiver filter bandwidths provides the best signal-to-noise ratio for SSB reception?

A. 500 Hz
B. 1000 Hz
C. 2400 Hz
D. 5000 Hz

For any type of phone (voice) reception, you want your receiver's bandwidth to accommodate all the voice frequencies transmitted, but no more than that. Excessive bandwidth allows more noise, as well as reducing the ability to reject adjacent signals. For best signal to noise ratio with Single Sideband (SSB), ***a bandwidth of about 2400 Hz is about right*** under most conditions. **ANSWER C.**

T4B08 What is the advantage of having multiple receive bandwidth choices on a multimode transceiver?

A. Permits monitoring several modes at once by selecting a separate filter for each mode
B. Permits noise or interference reduction by selecting a bandwidth matching the mode
C. Increases the number of frequencies that can be stored in memory
D. Increases the amount of offset between receive and transmit frequencies

On 6 and 10 meters SSB and CW, the worldwide radio offers bandwidth choices. ***When you change modes, your radio will automatically tighten up or expand the internal bandwidth filter selection.*** This allows you ***to minimize noise*** and maximize long range reception. **ANSWER B.**

T7A01 Which term describes the ability of a receiver to detect the presence of a signal?

A. Linearity
B. Sensitivity
C. Selectivity
D. Total Harmonic Distortion

The ability of a receiver to detect a weak radio signal is known as ***sensitivity***. For effective HF amateur radio operation, a receiver must have a power gain of approximately 10,000,000:1 (or about 70 decibels as measured at the antenna terminals) to bring a medium strength HF signal to a comfortable listening level with a set of earphones. As daunting as this task may seem, this sort of sensitivity is extremely easy to achieve with modern electronic components. In fact, a very primitive – but still popular – receiver known as a regenerative receiver can achieve this kind of performance with just a handful of components! Every ham should build a regenerative receiver sometime in his "career." Some of us have built lots of them! **ANSWER B.**

HAM HINT: *Your first radio should be a two-band 2 meter/70 cm FM handheld. The handheld keeps you in touch, locally. You also get some out-of-state and out-of-country DX, too, when local repeaters tie into the internet using voice over Internet protocol (VOIP).*

But, if you're really serious about exploring SSB satellite calls or moon bounce, meteor scatter, aurora, and 6 meter and 10 meter voice skip, plus all the fun of CW on high frequency Technician sub-bands, then consider a worldwide, high-frequency transceiver that includes 6 meters, 2 meters and 70 cm.

Not many HF transceivers give you these higher bands – sometimes they may give you 6 meters, but nothing higher. Each manufacturer has one or two specialty HF transceivers that give you the VHF and UHF bands, as well! They run about $900 to $1,400, but what you get is a worldwide DX transceiver, plus multimode capability on 6 meters, 2 meters, and 70 cm too! Best of all, everything is built into one nice neat, compact unit – nothing else to add other than your higher-band VHF and UHF antenna systems.

These are 12 volt radios that can be operated mobile as well as from a power supply at the house. This is a great way to have a single radio that will serve you well all the way up to your Extra Class license! Oh yeah, they give you 2 meter and 70 cm FM, too!

T7A03 Which of the following is used to convert a signal from one frequency to another?

A. Phase splitter
B. Mixer
C. Inverter
D. Amplifier

The ***mixer*** is a non-linear device that takes two radio signals and creates both the sum and the difference frequencies of the two input frequencies. A simple diode can act as a mixer, but most modern-day radios have a much more elegant arrangement using several diodes or active devices such as transistors or FETs to create a mixer circuit. Mixers are essential to nearly every modern radio design and are crucial to the overall performance of the radio in most cases. **ANSWER B.**

T7A04 Which term describes the ability of a receiver to discriminate between multiple signals?

A. Discrimination ratio
B. Sensitivity
C. Selectivity
D. Harmonic distortion

The capability of a radio to specifically hear one frequency and discriminate against signals on either side of that frequency is called ***selectivity***. On VHF and UHF equipment the selectivity is usually pre-set and most equipment is plenty selective. On the worldwide radio gear, you can sometimes add more filters for increased selectivity, improving noise reduction. While additional filters may cut down on the fidelity of the received signal, you can "tighten up" on the receiver response. One of the wonderful features of the new SDR (software defined) radios is that the receive bandwidth is continuously variable, from a "barn door" bandwidth of 8 kilohertz or so, to a razor thin 150 Hz or less. **ANSWER C.**

T7A11 Where is an RF preamplifier installed?

A. Between the antenna and receiver
B. At the output of the transmitter power amplifier
C. Between the transmitter and the antenna tuner
D. At the output of the receiver audio amplifier

Don't go out and buy one. Your radio equipment already has an excellent built-in preamplifier. Only if you purchased a very old radio, with noticeably weak reception, would you consider an external preamplifier. ***The pre-amp goes in between the antenna and the receiver***. Be careful how you wire it in – if you simply put it on the outside of the equipment, the first time you transmit, the pre-amp is history! It either needs external switching, or must be placed inside the equipment, between the antenna line in and the receiver section. The proper use of receiving preamplifiers is detailed in Eric's book ***Receiving Antennas for the Radio Amateur.*** **ANSWER A.**

Some solid state power amplifiers also incorporate built-in, sensitive receiver pre-amplifers as an option.

T7A06 What device converts the RF input and output of a transceiver to another band?

A. High-pass filter
B. Low-pass filter
C. Transverter
D. Phase converter

Many folks become ham radio operators from their experience on CB, 27 MHz. Some of their 27 MHz equipment may be repurposed and tied into a device called a "transverter," which converts the 11-meter signal to another band entirely. To reverse this process, you will need a receiving converter or downconverter, which takes another ham band signal and downconverts it to 27 MHz. While there are transverters that perform both functions in a single package, you can't assume that all "transverters" work on the receive side. So, "caveat emptor" when purchasing – or better yet, build your own! Transverters are great beginning homebrew projects! If you have some older CB radio, multimode, equipment don't toss it just yet. You might be able to use it on ham frequencies with a ***transverter***. **ANSWER C.**

Gordon uses a 10 GHz transverter that down-converts the received signal to 144 MHz into his weak-signal, multi-mode radio for maritime microwave communications.

T8D07 Which of the following describes DMR?

A. A technique for time-multiplexing two digital voice signals on a single 12.5 kHz repeater channel
B. An automatic position tracking mode for FM mobiles communicating through repeaters
C. An automatic computer logging technique for hands-off logging when communicating while operating a vehicle
D. A digital technique for transmitting on two repeater inputs simultaneously for automatic error correction

Digital Mobile Radio ***(DMR)*** was originally developed for commercial radio use but is finding increasing acceptance in the amateur radio community. ***Time multiplexing***, the sharing of a channel by "chopping" the signal into time slots, has been around for a long time in the telephone business, and has been adapted by hams with minimal change from its commercial version. **ANSWER A.**

Digital handheld radios with their talk groups provide crystal clear contacts with no background noise.
Photo courtesy of BridgeCom

DIGIPEATER

A "traditional" VHF or UHF analog repeater is a complicated, finely-tuned device. One of the biggest challenges to making a repeater work effectively is totally separating the input and output signals. Any "leakage" from one signal path to the other can seriously degrade repeater performance, or even making it utterly non-functional. While the normal frequency "split" used on a repeater is one step toward solving a difficult engineering problem, it is only the first of many issues the repeater technician needs to overcome. Because of space limitations, most repeaters must use the same antenna for transmitting and receiving. The device that allows both the receiver and the transmitter to use the same antenna at the same time is called a duplexer, and is usually the most expensive item in any repeater installation, and understandably so, when you realize what the duplexer actually has to accomplish!

One of the major (and amazing) advantages of many types of digital radio signals is that they can be "digipeated" or relayed through a simplex repeater, which is little more than an FM transceiver. Since the input and output signals of a digipeater do not occur at the same time, the entire duplexer problem is neatly bypassed. The first use of digipeaters was with packet radio, which had its Golden Age in the early 1990s. Anyone with a 2 meter radio, an inexpensive Terminal Node Controller (TNC), and a ground plane antenna could put up a very effective digipeater! A quarter of a century later, the tried and true digipeater concept is being applied to voice transmissions in a number of intriguing formats.

By the way, plain "vanilla" packet radio is seeing a bit of a renaissance in many areas of the country because of its simplicity and vast supply of inexpensive surplus equipment from the Golden Age of packet! It is a very effective backup system for short emergency messages and such.

T2B12 What is the purpose of the color code used on DMR repeater systems?

A. Must match the repeater color code for access
B. Defines the frequency pair to use
C. Identifies the codec used
D. Defines the minimum signal level required for access

Digital mobile radio (DRM) is just one of the many digital radio systems. ***When the color codes used by DRM match the repeater, access is granted.*** They are the digital equivalent of the CTCSS codes used by FM repeaters. **ANSWER A.**

T2B07 How can you join a digital repeater's "talkgroup"?

A. Register your radio with the local FCC office
B. Join the repeater owner's club
C. Program your radio with the group's ID or code
D. Sign your call after the courtesy tone

One of the unique capabilities of digital audio is that it can be "routed" or "groomed" on a packet-by-packet basis. The voice packets can, therefore, be filtered according to some useful parameter, such as an interest group. ***Programming the group ID*** on your radio will route only those packets associated with your group to reach your radio (and vice versa). **ANSWER C.**

T8D02 What is a "talkgroup" on a DMR repeater?

A. A group of operators sharing common interests
B. A way for groups of users to share a channel at different times without hearing other users on the channel
C. A protocol that increases the signal-to-noise ratio when multiple repeaters are linked together
D. A net that meets at a specified time

With digital radio, multiple groups can use the same carrier frequency at the same time by routing the voice packets into different groups. A talk group is one with the same group ID programmed into their radios. They can blissfully ***operate without any awareness of other talk groups using the same carrier frequency***. Pretty amazing stuff, actually! **ANSWER B.**

T4B07 What does a DMR "code plug" contain?

A. Your call sign in CW for automatic identification
B. Access information for repeaters and talkgroups
C. The codec for digitizing audio
D. The DMR software version

A code plug is not a physical plug like a thumb drive, but simply a block of programming that has ***information about local repeaters and talk groups***, so you can get your DMR transceiver up and running quickly. **ANSWER B.**

T4B09 How is a specific group of stations selected on a digital voice transceiver?

A. By retrieving the frequencies from transceiver memory
B. By enabling the group's CTCSS tone
C. By entering the group's identification code
D. By activating automatic identification

DMR repeaters are far more selective than, say, the CTCSS tones on a "standard" repeater. You can control access on a person-by-person basis on a DMR repeater by requiring the user to have a ***specific group identification code***. **ANSWER C.**

T4B11 Which of the following must be programmed into a D-STAR digital transceiver before transmitting?

A. Your call sign
B. Your output power
C. The codec type being used
D. All these choices are correct

Since a D-STAR transmitter is available through the internet, there has to be some means of preventing non-hams from transmitting. Requiring a ***valid call sign*** to be programmed into the rig before it can transmit is just the first line of defense against unauthorized use. **ANSWER A.**

T8D12 Which of the following best describes an amateur radio mesh network?

A. An amateur-radio based data network using commercial Wi-Fi equipment with modified firmware
B. A wide-bandwidth digital voice mode employing DMR protocols
C. A satellite communications network using modified commercial satellite TV hardware
D. An internet linking protocol used to network repeaters

Hams are cheap! If there's a way some readily available, inexpensive technology can be pressed into some exotic new application, some ham or group of hams have figured out how to do it. The use of modified consumer Wi-Fi equipment has allowed hams (at least in some regions) to build a shadow ***amateur-radio-based data network*** totally independent of the commercial infrastructure. Radio amateurs have twelve microwave bands that are almost unused, great playgrounds where we can experiment with more of this broadband technology. **ANSWER A.**

Time Division Multiplexing

Several amateur radio modes use the principle of Time Division Multiplexing (TDM) to allow several conversations to take place at the same time on the same frequency. A conversation is "chopped up" into very small segments of time with gaps in between, like the teeth of a comb. A second conversation chopped up in the same manner can be fit into the space between the "teeth" of the first conversation. DMR (Digital Mobile Radio) allows two conversations to be interlaced in this manner.

Full TDMA (time division multiple access) allows multiple conversations to be interlaced in the same manner. You choose which conversation you want to participate in by choosing the particular "set of teeth" or time slot.

This principle of time division multiplexing has been used for the telephone system for more than a half century, but it's only been catching on recently in Amateur Radio.

Website Resources

▼ IF YOU'RE LOOKING FOR	▼ THEN VISIT
A great rig to get your feet wet on HF. Excellent rig for reliable mobile operation as well.	https://www.dxengineering.com/parts/alo-dx-10
Every radio amateur should have at least one general coverage receiver in the shack.	https://swling.com/blog/2014/06/the-best-general-coverage-transceivers-for-shortwave-listening/
Great app for decoding NAVTEX, SITOR and AMTOR transmissions. A must for the seafaring ham.	http://www.wolphi.com/marine-apps/droidnavtex/
Hams should always know where they are! This application will tell you which GPS satellites are available for your particular QTH.	https://gps-test.droidinformer.org/

▼ IF YOU'RE LOOKING FOR	▼ THEN VISIT
Here is a list of no less than 99 additional ham radio 'droid applications, and reviews by a lot of hams.	http://www.cqdx.ru/ham/new-equipment/ham-radio-android-apps-list/
Using DMR	https://youtu.be/5FAFt1QCtC0
DSP Speakers	www.BHI-Ltd.com
HF radio/ amplifier	www.Elecraft.com
HF radio	www.Flex-Radio.com
Auto Tuners	www.LDGElectronics.com
Radio DSP	www.WestMountainRadiocom

Watch This! **https://www.youtube.com/playlist?list=PL1KAjn5rGhizk5f2_whBLJaOHj59OCl_3 (video 83)**

Run Some Interference Protection

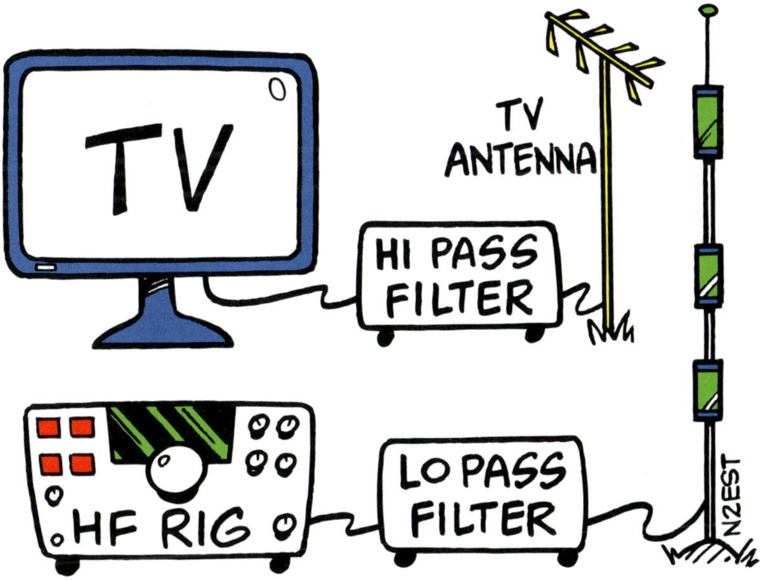

T7B10 What might be a problem if you receive a report that your audio signal through an FM repeater is distorted or unintelligible?

A. Your transmitter is slightly off frequency
B. Your batteries are running low
C. You are in a bad location
D. All these choices are correct

It seems that with radios there can be several causes for poor signal audio. Your new handheld could be slightly off frequency. You'll want to check back with your dealer or the manufacturer if this is suspected. You can expect reduced performance when the batteries begin to run low. But remember, don't transmit when connected to the wall charger. A bad location is the easiest to fix. Often moving just slightly can put you in a better place for the signal to be heard. You will find that ***all of these choices are correct*** at one time or another. VHF/UHF handheld radios have become so inexpensive that it's quite cost effective to have an extra one just for monitoring the quality of your transmitted signal. This is a surefire method of assuring that you aren't overmodulating or otherwise generating a sub-standard signal. **ANSWER D.**

T4B01 What is the effect of excessive microphone gain on SSB transmissions?

A. Frequency instability
B. Distorted transmitted audio
C. Increased SWR
D. All these choices are correct

Some base station radios for high frequency 10-meter work, and for weak signal VHF and UHF operation, have a microphone gain control. Set the control at about half scale. Turning the control wide open will set the ***mic gain too high*** and cause the transmitted ***signal*** to become ***distorted***. **ANSWER B.**

T7B01 What can you do if you are told your FM handheld or mobile transceiver is over-deviating?

A. Talk louder into the microphone
B. Let the transceiver cool off
C. Change to a higher power level
D. Talk farther away from the microphone

If your set is over-deviating it means that too much modulation is driving your signal beyond its normal bandwidth. If you ***talk farther away from the microphone*** you will minimize or even eliminate the over-deviation. This should be considered a temporary fix. If your rig is consistently over-deviating, you should have the modulation circuits adjusted, preferably using a reliable communications monitor. **ANSWER D.**

T2B05 What would cause your FM transmission audio to be distorted on voice peaks?

A. Your repeater offset is inverted
B. You need to talk louder
C. You are talking too loudly
D. Your transmit power is too high

One thing you can say about FM: when it's good it's really, really good, and when it's bad, it's really, really bad. While slightly over modulating an AM or SSB transmitter may cause some annoying distortion, over modulating (over deviating) an FM transmitter causes the signal to break up on syllable peaks, especially on narrow-band repeaters. As a temporary fix for over deviation, simply ***talk more quietly or move your mouth a little farther from your microphone*** until people can understand you. Also, be sure your radio is set for narrow band operation; many imported radios are set by default to wideband operation, a sure-fire recipe for extreme overmodulation on a narrowband receiver! **ANSWER C.**

T4B05 What does the scanning function of an FM transceiver do?

A. Checks incoming signal deviation
B. Prevents interference to nearby repeaters
C. Tunes through a range of frequencies to check for activity
D. Checks for messages left on a digital bulletin board

Most modern VHF transceivers have a scanning function, which can range from fairly elementary to quite elaborate. Amateur "scanners" can cover frequencies both within and outside the amateur bands. Probably the most common and useful "outside" frequency is the National Weather Service on 162.550 MHz. It used to be quite common to also program local emergency services such as police and fire frequencies into scanners, but these are almost all encrypted nowadays, so all you get is noise. But being able to ***scan all your local repeaters and popular simplex frequencies for activity*** is almost a "must." **ANSWER C.**

T7B09 What should be the first step to resolve non-fiber optic cable TV interference caused by your amateur radio transmission?

A. Add a low-pass filter to the TV antenna input
B. Add a high-pass filter to the TV antenna input
C. Add a preamplifier to the TV antenna input
D. Be sure all TV feed line coaxial connectors are installed properly

In any interference case, you want to be sure you have a "clean house" to start with. Assuming you are transmitting a clean signal on your 10 meter radio, this situation indicates cable leakage, which can result from any one of possibly dozens of ***improperly installed cable connectors*** in a cable TV installation. Never install any type of filter on cable or satellite TV connections. Just make sure all connections are finger tight. **ANSWER D.**

T7B03 Which of the following can cause radio frequency interference?

A. Fundamental overload
B. Harmonics
C. Spurious emissions
D. All these choices are correct

You are outside sitting in your deck chair listening to your local repeater when suddenly your handheld picks up nothing but hash. Your buddy across the street (also a ham) is transmitting with a mobile radio and amplifier that he bought off a swap net that was advertised as "may need some tweaking." That nearby transmitter is probably ***overloading*** your receiver with ***harmonics and spurious emissions***, and ***all of this leads to radio frequency interference***. Don't buy questionable used gear – Marconi himself probably couldn't fix it. **ANSWER D.**

T7B11 What is a symptom of RF feedback in a transmitter or transceiver?

A. Excessive SWR at the antenna connection
B. The transmitter will not stay on the desired frequency
C. Reports of garbled, distorted, or unintelligible voice transmissions
D. Frequent blowing of power supply fuses

You are on 6 meters for the first time, and signal strength tests with hams a few miles away indicate you have a strong signal, but your modulation ***(your voice) is slightly garbled***. This is most likely ***caused by RF feedback*** between your antenna system and the microphone. You have the antenna temporarily set on a tripod, just a few feet away from your operating station. Get some longer coax, move the antenna at least 30 feet away from your operating station, and likely your friends will tell you the distortion has cleared up. No more RF feedback! **ANSWER C.**

T6D03 Which of the following is a reason to use shielded wire?

A. To decrease the resistance of DC power connections
B. To increase the current carrying capability of the wire
C. To prevent coupling of unwanted signals to or from the wire
D. To couple the wire to other signals

All electrical conductors are capable of both radiating and receiving radio frequency energy. This is great in the case of an antenna; it's not so great in most other applications. When ***you don't want a conductor to radiate or receive radio signals*** (or other interference), the use of shielding is often an effective solution. The most familiar ***shielded wire*** is coaxial cable. **ANSWER C.**

T7B06 Which of the following actions should you take if a neighbor tells you that your station's transmissions are interfering with their radio or TV reception?

A. Make sure that your station is functioning properly and that it does not cause interference to your own radio or television when it is tuned to the same channel
B. Immediately turn off your transmitter and contact the nearest FCC office for assistance
C. Install a harmonic doubler on the output of your transmitter and tune it until the interference is eliminated
D. All these choices are correct

First, double check that your equipment is well grounded and operating properly, and ***double check that your TV is working okay*** when you are transmitting over the air. If it is, chances are your neighbors may have some loose connections on their TV receivers, and this should be cured as a step toward eliminating the interference from your station. **ANSWER A.**

T7B02 What would cause a broadcast AM or FM radio to receive an amateur radio transmission unintentionally?

A. The receiver is unable to reject strong signals outside the AM or FM band
B. The microphone gain of the transmitter is turned up too high
C. The audio amplifier of the transmitter is overloaded
D. The deviation of an FM transmitter is set too low

Interference between Amateur Radio and other radio services has always been a potential problem, though in many regards it's not as severe as it once was. As broadcast TV signals changed from analog to digital, the "disappearance" of Channel 2 TV, for instance, has made full power 6-meter amateur radio a joy in recent years, where in the past it was almost certain to cause interference to nearby analog TVs on that channel. In any case, it behooves any amateur to understand some basics of preventing or curing various forms of RF interference as they might crop up. Often, however, ***the problem is design defects in the consumer electronic device***. A small degree of wisdom (or a lot of experience) is often necessary to resolve these issues with your neighbor. Each case is a special case. **ANSWER A.**

T7B04 Which of the following could you use to cure distorted audio caused by RF current on the shield of a microphone cable?

A. Band-pass filter
B. Low-pass filter
C. Preamplifier
D. Ferrite choke

RF (radio frequency) current flowing on the outside of microphone cables (and other conductors as well) is generally of the common mode sort, which is normally most effectively cured by the use of ferrite beads or cores surrounding the conductor, thus "choking" off the RF current. Sometimes reducing RF problems is a bit of an art form, but always start with the "standard" fixes first. If you plan to go digital with your new Technician Class privileges, load up on a handful of ***ferrite chokes***. These are clamshell iron devices that simply snap on over the wiring coming from your handheld or mobile radio to your computer. The ferrite choke will minimize RF currents flowing on the shield of an audio cable. You want everything on the inside, not the outside! **ANSWER D.**

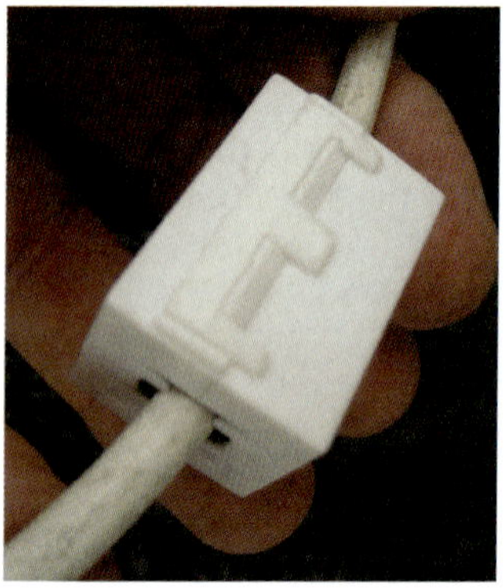

Snap-on ferrite chokes can help solve interference problems. Visit: Palomar-Engineers.com

---

T7B05 How can fundamental overload of a non-amateur radio or TV receiver by an amateur signal be reduced or eliminated?

A. Block the amateur signal with a filter at the antenna input of the affected receiver
B. Block the interfering signal with a filter on the amateur transmitter
C. Switch the transmitter from FM to SSB
D. Switch the transmitter to a narrow-band mode

Unfortunately, a lot of consumer equipment is designed and built as if they were the only occupants of the radio spectrum. No matter how pristine and clean your amateur radio signal is, some types of interference, such as front-end overload, can only be fixed at the consumer's equipment. This is just one such case. Consult the ARRL Handbook or other reputable references for detailed information on how to ***apply filters to the consumer gear*** for this solution. **ANSWER A.**

T7B07 Which of the following can reduce overload of a VHF transceiver by a nearby commercial FM station?

A. Installing an RF preamplifier
B. Using double-shielded coaxial cable
C. Installing bypass capacitors on the microphone cable
D. Installing a band-reject filter

A band-reject filter removes (or at least greatly reduces) a specific range of radio frequencies. To prevent overloading of a sensitive VHF receiver "front end" by a local high-power FM broadcast station, it is often useful to reject the entire FM broadcast band ***by means of a band-reject filter***. Under overload conditions, a wideband RF preamplifier is the last thing you want, as it will generally exacerbate the problem! **ANSWER D.**

T7B08 What should you do if something in a neighbor's home is causing harmful interference to your amateur station?

A. Work with your neighbor to identify the offending device
B. Politely inform your neighbor that FCC rules prohibit the use of devices that cause interference
C. Make sure your station meets the standards of good amateur practice
D. All these choices are correct

Encountering radio interference from your neighbor is rare on 2 meters or 70cm operation. Nonetheless, the modern microprocessor circuits in home electronics can do some interesting things over your radio receiver! One neighbor has a washing machine that on the spin cycle creates buzzing on my 6 meter radio. Another neighbor just put in a new driveway alarm system, and when they drive in I hear a phantom bing-bong radio signal over the 10 meter band. Work with your neighbor to try and resolve the problem of interference to your ham station. Explain to your neighbor your knowledge of the radio spectrum, and together try and work out a solution. Also, never ignore a neighbor who may claim that your radio's transmissions are interfering with their home electronics. Most interference can be resolved by hams and neighbors cooperating to solve the issue by first identifying the source and taking steps to resolve the interference. ***All of the answers to this question are correct***. **ANSWER D.**

Simple snap-on choke filters like these can help resolve harmful interference problems on Part 15 devices.

Ham Hint: *Never put filters on satellite or cable TV coax cable connections. Only a qualified cable or satellite TV antenna technician should be working on the coaxial cable coming into a TV set. The very best thing you can do is to make sure all TV antenna F connectors are finger tight. A loose F connector will sometimes create interference.*

Website Resources

IF YOU'RE LOOKING FOR	THEN VISIT
Multi-mode weak signal modes	www.arrl.org/digital-data-modes www.ac6v.com/opmodes.htm www.hfradio.org.uk
Licking RFI interference	www.arrl.org/radio-frequency-interference-rfi
DXZONE.com	www.tinyurl.com/nox6o
RF chokes	www.Palomar-Engineers.com

Watch This! https://www.youtube.com/playlist?list=PL1KAjn5rGhizk5f2_whBLJaOHj59OCl_3 (video 80)

CD 3 TRACK 6

Electrons, Go With The Flow

T5A05 What is the electrical term for the force that causes electron flow?

A. Voltage
B. Ampere-hours
C. Capacitance
D. Inductance

Think of voltage as water pressure in your kitchen plumbing! Open the valve (turn on the switch) and current begins to flow. The pressure is ***voltage***, and the trickle of water is similar to current. **ANSWER A.**

T7D01 Which instrument would you use to measure electric potential?

A. An ammeter
B. A voltmeter
C. A wavemeter
D. An ohmmeter

Another name for electromotive force is voltage. We measure voltage with a ***voltmeter***. **ANSWER B.**

Meter displaying volts and amps.

T7D02 How is a voltmeter connected to a component to measure applied voltage?

A. In series
B. In parallel
C. In quadrature
D. In phase

We test for voltage by hooking our meter across the voltage source without undoing any wires – a ***parallel connection***. Checking the voltage when operating equipment is called "checking the voltage source under load." **ANSWER B.**

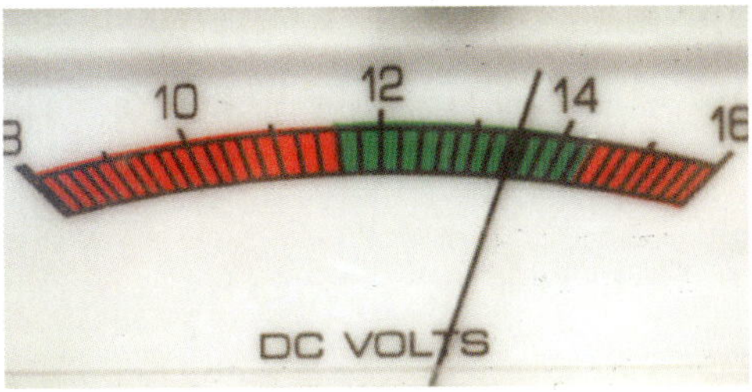

Meter displaying about 13.5 volts.

T6A10 Which of the following battery chemistries is rechargeable?

A. Nickel-metal hydride
B. Lithium-ion
C. Lead-acid
D. All these choices are correct

Rechargeable batteries are sometimes known as "secondary cells." Many new, wonderful types of rechargeable batteries make portable and emergency amateur radio operation more practical than ever. NMh (Nickel Metal Hydride) is common in HTs, and Li-Ion (Lithium Ion) is found in step up HT battery systems. ***All of these*** rechargeable batteries use different types of chargers. Just be sure you use the right battery for your particular device and the right charger for that battery! And follow the safety warnings! Modern batteries are capable of energy levels unimaginable for some of us old timers! **ANSWER D.**

Ni-Cad rechargeable 1.25 volt batteries in a marine handheld

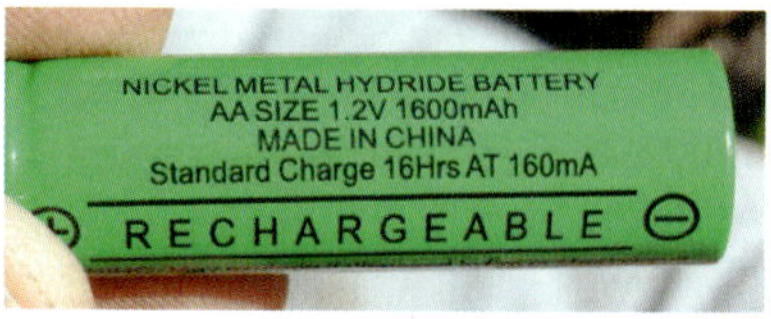

If a battery is rechargeable it will say so.

T6A11 Which of the following battery chemistries is not rechargeable?

A. Nickel-cadmium
B. Carbon-zinc
C. Lead-acid
D. Lithium-ion

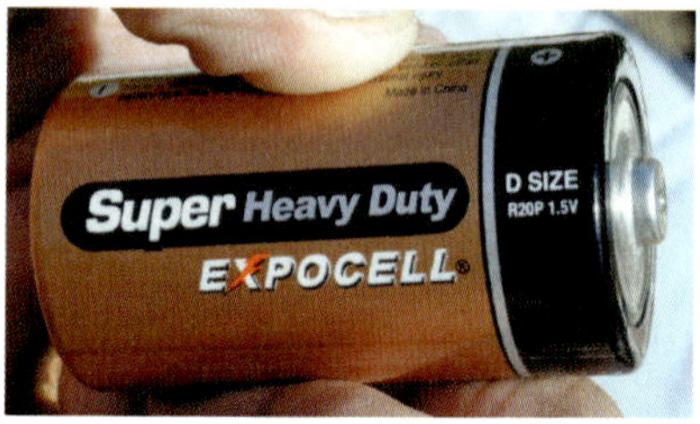

Luckily you won't find any handheld shipped with ***carbon-zinc, non-rechargeable batteries***. Alkaline and carbon-zinc batteries only work once and should be properly disposed of when depleted. These individual cells make up a good battery source for HTs that come with a battery tray. The battery trays have no contacts on the bottom to accidentally put them in a recharge device. Alkaline and carbon-zinc cells must never be recharged. **ANSWER B.**

T4A09 How can you determine the length of time that equipment can be powered from a battery?

A. Divide the watt-hour rating of the battery by the peak power consumption of the equipment
B. Divide the battery ampere-hour rating by the average current draw of the equipment
C. Multiply the watts per hour consumed by the equipment by the battery power rating
D. Multiply the square of the current rating of the battery by the input resistance of the equipment

The capacity of a battery is measured in amp-hours, which is amps times hours. ***To get hours, you simply divide amp-hours by amps.*** **ANSWER B.**

T4A11 Where should the negative power return of a mobile transceiver be connected in a vehicle?

A. At the 12 volt battery chassis ground
B. At the antenna mount
C. To any metal part of the vehicle
D. Through the transceiver's mounting bracket

Ham radio power leads need to be connected directly at the battery source. This means positive to the battery positive terminal, and ***negative to the battery negative terminal or to the nearby engine block ground strap***. This will minimize alternator whines and whistles! Also be sure the power cable is FUSED at the battery! Automotive batteries are capable of supplying tremendous current, and the last thing you want to do is create an automotive meltdown as a result of a malfunctioning or improperly installed mobile ham station! **ANSWER A.**

T5A03 What is the name for the flow of electrons in an electric circuit?

A. Voltage
B. Resistance
C. Capacitance
D. Current

Think of the flow of electrons as the flow of water in a stream. If you get out there in midstream, you will feel the ***current***. **ANSWER D.**

T7D04 Which instrument is used to measure electric current?

A. An ohmmeter
B. An electrometer
C. A voltmeter
D. An ammeter

Current is measured in amperes, the unit of current. We use an ***ammeter to measure electrical current***. **ANSWER D.**

T5A01 Electrical current is measured in which of the following units?

A. Volts
B. Watts
C. Ohms
D. Amperes

Current is measured in ***amperes***. Amperes is often referred to as "amps." **ANSWER D.**

T5A07 Why are metals generally good conductors of electricity?

A. They have relatively high density
B. They have many free electrons
C. They have many free protons
D. All these choices are correct

It is called *electronics* because it is electrons that do all the work! In any conductor, it's the movement of electrons that allows things to happen, and ***the more free electrons there are, the better the conductor conducts.*** Most metals have a lot of free electrons, though there are great differences in the conductivity of different metals. **ANSWER B.**

T5A09 Which of the following describes alternating current?

A. Current that alternates between a positive direction and zero
B. Current that alternates between a negative direction and zero
C. Current that alternates between positive and negative directions
D. All these ANSWERs are correct

Have you ever been bitten by a hot power cord? Think back to that shocking time and recall your accidental zap feeling much like a buzz. It is exactly that – ***alternating current*** reversing directions 60 times a second. **ANSWER C.**

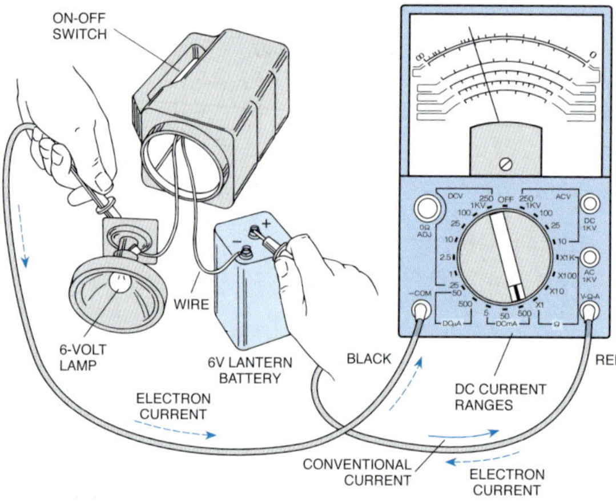

Multimeter to Measure a Series Circuit

Source: *Basic Electronics* © 1994, 2000, Master Publishing, Inc., Glencoe, Illinois

T7D03 When configured to measure current, how is a multimeter connected to a component?

A. In series
B. In parallel
C. In quadrature
D. In phase

An ammeter measures current. To measure current, turn off the power, disconnect one lead of the load from its source voltage (for example, at the fuse holder) and insert an ***ammeter in series*** with that lead. If you are measuring DC current, you will need to connect the meter with the correct polarity so the meter reads up scale when power is turned on. (Never connect an ammeter directly across a voltage source! You will either fry your meter or blow a rather expensive fuse. In every electronics class I've ever taught, two or three students just have to try this out and see for themselves!) An ammeter in parallel appears as a dead short to any voltage source. An ammeter always connects in series, never across/parallel to a voltage source. **ANSWER A.**

T4A03 Why are short, heavy-gauge wires used for a transceiver's DC power connection?

A. To minimize voltage drop when transmitting
B. To provide a good counterpoise for the antenna
C. To avoid RF interference
D. All these choices are correct

All wire has some amount of resistance in it. This resistance will cause the voltage to drop proportionally through the wire, according to Ohm's Law. The more current you draw, the more the voltage drop. ***Short, fat wires have much less resistance than long, skinny ones, so they will have less voltage drop***. You should always measure any DC power supply voltages at the LOAD (the radio) end of any cables, while under normal operating conditions, to ensure that you have ***enough voltage for proper operation***. **ANSWER A.**

T4A01 Which of the following is an appropriate power supply rating for a typical 50 watt output mobile FM transceiver?

A. 24.0 volts at 4 amperes
B. 13.8 volts at 4 amperes
C. 24.0 volts at 12 amperes
D. 13.8 volts at 12 amperes

A typical FM transceiver is *about* 30% efficient, which means in order to put out 50 watts, you'd need at *about* 170 watts of input power. We can immediately eliminate answers A and C because car batteries run at 13.8 volts (Nominal 12 volts). ***Input power is voltage times amps, so 13.8V x 12A =165.6 watts, which will just about do the job. If you know the input power, and you need to figure out the current, simply divide the input power by the voltage, in this case, 165.6W ÷ 13.8V = 12A.*** This is a considerable amount of current, however, and you do *not* want to power your mobile rig from a cigarette lighter or accessory socket. It should be wired directly to the battery with a high current fuse in both the positive and negative leads. **ANSWER D.**

T6D01 Which of the following devices or circuits changes an alternating current into a varying direct current signal?

A. Transformer
B. Rectifier
C. Amplifier
D. Reflector

Most ham equipment works off of DC voltage. The way we transform household AC power into DC is with a rectifier circuit. The rectifier uses diodes to block the full house power AC cycle, leading to a varying direct current signal. Capacitors are used to filter and smooth out this fluctuating DC current. Thanks to the ***rectifier*** circuit, we can change AC into varying DC. **ANSWER B.**

T6B02 What electronic component allows current to flow in only one direction?

A. Resistor
B. Fuse
C. Diode
D. Driven element

The process of changing AC to pulsating DC is called rectification. Big word, huh? We use a diode in a circuit to allow current to flow in only one direction. The ***diode*** – much like a check valve – stops current flow when it tries to go in the reverse direction. **ANSWER C.**

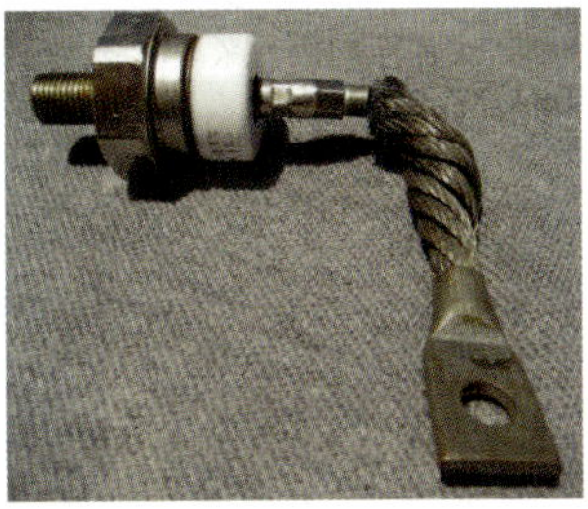

This large diode could protect a 12 volt DC linear amplifier.

T6B01 Which is true about forward voltage drop in a diode?

A. It is lower in some diode types than in others
B. It is proportional to peak inverse voltage
C. It indicates that the diode is defective
D. It has no impact on the voltage delivered to the load

A diode has a forward voltage drop that is *relatively* independent of the current passing through it, unlike a resistor where the voltage and current are exactly proportional. ***The actual voltage drop depends on the type of diode.*** A silicon diode has about a 0.6 volt forward voltage drop, a germanium diode has about a 0.25 volt forward drop, and a Schottky diode has just under 1 volt forward drop. **ANSWER A.**

T6B09 What are the names for the electrodes of a diode?

A. Plus and minus
B. Source and drain
C. Anode and cathode
D. Gate and base

First take a look at the symbol for a diode. The cathode is represented by the short straight line at the tip of the arrow and the anode is represented by the arrow itself. Now think backwards – current flows in the opposite direction of the arrow from cathode to ***anode***, forward bias. There is almost no current flow in the direction of the arrow from anode to ***cathode***. In a silicon diode, it takes about half a volt for conduction. **ANSWER C.**

T6B06 How is the cathode lead of a semiconductor diode often marked on the package?

A. With the word "cathode"
B. With a stripe
C. With the letter C
D. With the letter K

Not much room on the tiny diode to put a letter or a word to indicate the cathode end, so a ***tiny stripe*** does the trick nicely. This is the same reason a rancher uses a brand instead of a paragraph to identify his cow. **ANSWER B.**

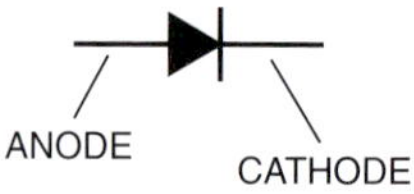

Here is the schematic symbol of a diode. Current will only flow ONE WAY in a diode. You can remember this diode diagram as a one-way arrow (key words).

Semiconductor Diode

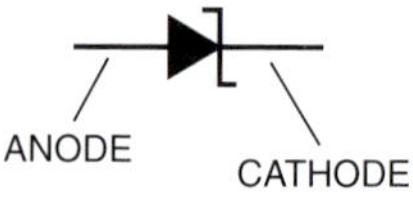

Here is the schematic symbol of a Zener diode. Since a diode only passes energy in one direction, look for that one-way arrow, plus a "Z" indicating it is a Zener diode. Doesn't that vertical line look like a tiny "Z"?

Zener Diode

T6A01 What electrical component opposes the flow of current in a DC circuit?

A. Inductor
B Resistor
C. Inverter
D. Transformer

The opposition to the flow of current in a direct current circuit is called resistance. The component is called a ***resistor***. Just like a beaver dam resists the current flow in a stream. **ANSWER B.**

Resistors.

T5A11 What type of current flow is opposed by resistance?

A. Direct current
B. Alternating current
C. RF current
D. All these choices are correct

By definition, ***resistance is the opposition of current flow of any kind***. However, resistance is not the only thing that will oppose current flow; reactance opposes alternating currents (of which RF is just one "flavor"). **ANSWER D.**

T5A04 What are the units of electrical resistance?

A. Siemens
B. Mhos
C. Ohms
D. Coulombs

The Ohm is the fundamental unit of electrical resistance, and one of the three components of Ohm's Law: E = I x R. It is the measure of how much a component opposes the flow of electrical current. The reciprocal of the Ohm is the Mho, which is Ohm spelled backwards. It is a measure of how well a component *conducts* electric current. It's easy to remember, because with mo' mho, you have mo' current flow. What could be simpler than that? **ANSWER C.**

T6A02 What type of component is often used as an adjustable volume control?

A. Fixed resistor
B. Power resistor
C. Potentiometer
D. Transformer

Ham operators are about the last to enjoy the potentiometer. This is like an old-fashioned variable resistor where a wiper brush makes contact with wire windings, varying the resistance. This is found in the volume control of most handheld ham radios. That new plasma TV doesn't have a potentiometer; volume is controlled by a digital push button circuit. In ham sets, we usually have the beloved ***potentiometer!*** **ANSWER C.**

T6A03 What electrical parameter is controlled by a potentiometer?

A. Inductance
B. Resistance
C. Capacitance
D. Field strength

The ***potentiometer*** varies the ***resistance*** in most ham radio volume control circuits. If your ham set does not have a round volume control knob, but only push buttons for volume control, you don't have a potentiometer. **ANSWER B.**

T5A08 Which of the following is a good electrical insulator?

A. Copper
B. Glass
C. Aluminum
D. Mercury

Old-time hams will lead you down to their basement to show off their collection of multi-colored ***glass*** power pole insulators. Clean glass is a great insulator, and generally won't conduct electrons unless it gets covered with dirt and ocean salt air. Power pole maintenance workers use a special non-conductive water jet to clean those large, high-voltage insulators. **ANSWER B.**

T6A06 What type of electrical component stores energy in a magnetic field?

A. Varistor
B. Capacitor
C. Inductor
D. Diode

It is the coil, called an inductor, that stores energy in the magnetic field. We can actually see the effects of an ***inductor*** with a magnetic compass. **ANSWER C.**

Inductor.

T6A07 What electrical component is typically constructed as a coil of wire?

A. Switch
B. Capacitor
C. Diode
D. Inductor

The coil of wire is found in the electrical component called an ***inductor***. **ANSWER D.**

Variable inductors.

T5C03 What describes the ability to store energy in a magnetic field?

A. Admittance
B. Capacitance
C. Resistance
D. Inductance

When exposed to current, coils develop a magnetic field, which can be detected by holding a magnetic compass near the energized coil. Energy stored in a magnetic field is called ***inductance***. **ANSWER D.**

T5C04 What is the unit of inductance?

A. The coulomb
B. The farad
C. The henry
D. The ohm

The basic unit of ***inductance is the henry***. Because the henry is a fairly large unit, we usually measure inductance in one thousandths of a henry (millihenry) or one millionths of a henry (microhenry). By the way, the plural of henry is not "henries" but "henrys." **ANSWER C.**

T5C12 What is impedance?

A. The opposition to AC current flow
B. The inverse of resistance
C. The Q or Quality Factor of a component
D. The power handling capability of a component

Impedance is probably the most important concept to understand in all of radio. It is ***the composite opposition to the flow of current in an AC (or radio frequency) circuit***. It is sometimes erroneously called "AC resistance" but this is an incomplete description. Impedance is the result of both resistance and reactance and is dependent on frequency. If you can consistently work out impedance problems, you can safely say you know some electronics! **ANSWER A.**

T5C05 What is the unit of impedance?

A. The volt
B. The ampere
C. The coulomb
D. The ohm

Impedance, represented by the letter Z, is the total ohmic special resistance that opposes current flow within an alternating current circuit. Even though impedance is a complex number, it follows Ohm's Law, which is fortunate. The unit of ***impedance***, just as resistance, ***is the Ohm***. Ohm's law for impedance is E = I x Z. **ANSWER D.**

T5C01 What describes the ability to store energy in an electric field?

A. Inductance
B. Resistance
C. Tolerance
D. Capacitance

Capacitors store energy in an electric field. This is called ***capacitance***. The earlier questions asked about inductance, which stores energy in a magnetic field. Remember, the magnetic and electric fields exist at right angles to each other in all radio waves. **ANSWER D.**

T5C02 What is the unit of capacitance?

A. The farad
B. The ohm
C. The volt
D. The henry

The basic unit of ***capacitance*** is the ***farad***. Because the farad is a fairly large unit, we usually measure capacitance in one millionths of a farad (microfarad) or one million millionths of a farad (picofarad). **ANSWER A.**

T6A04 What electrical component stores energy in an electric field?

A. Varistor
B. Capacitor
C. Inductor
D. Diode

It is the capacitor that stores energy in an electric field. Got it? The electric field has capacitance which is stored in a ***capacitor***. **ANSWER B.**

T6A05 What type of electrical component consists of conductive surfaces separated by an insulator?

A. Resistor
B. Potentiometer
C. Oscillator
D. Capacitor

The ***capacitor has 2 or more conductive surfaces, separated by insulation***. The insulator may be air, vacuum, or a solid material such as plastic or paper. When a capacitor shorts out or breaks down, it's usually one where the insulation has finally dried up and failed. This results in all sorts of noise as it becomes a leaky older capacitor. **ANSWER D.**

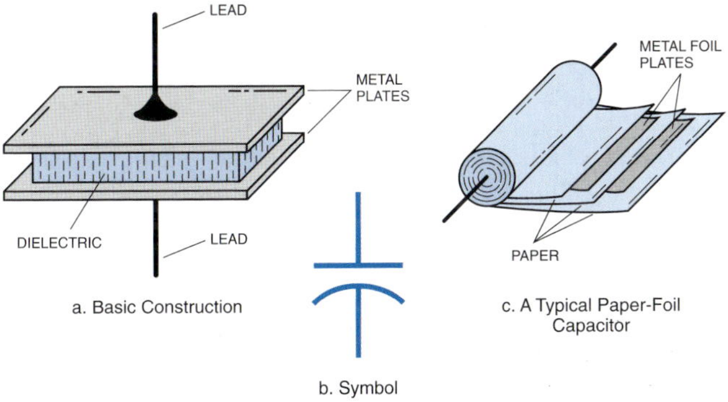

Typical construction and schematic symbol for capacitors.

T6A09 What electrical component is used to protect other circuit components from current overloads?

A. Fuse
B. Thyratron
C. Varactor
D. All these choices are correct

We use a ***fuse*** to protect components from excessive amounts of current flow. Even handheld radio batteries have fuses, in case you take that freshly charged battery and stick it in your pocket with a lot of spare change. The battery shorts out, you get a burning sensation in your thigh, but luckily the fuse opens, and all you get is a dead battery and a lot of hot quarters. Some batteries use thermal fuses, which automatically reset once they cool down. Always be sure to replace a ***fuse*** with one of the proper rating. **ANSWER A.**

TRANSISTOR BASICS

The transistor is the fundamental building block of modern electronic devices, and is ubiquitous in modern electronic systems. Prior to the development of transistors, vacuum (electron) tubes were the main active components in electronic equipment.

Following its development in 1947 by John Bardeen, Walter Brattain, and William Shockley, the transistor revolutionized the field of electronics, and paved the way for smaller and cheaper radios, calculators, and computers, among other things. The transistor is on the list of IEEE milestones in electronics, and the inventors were jointly awarded the 1956 Nobel Prize in Physics for their achievement.

A transistor is a semiconductor device used to amplify and switch electronic signals and electrical power. It is composed of semiconductor material with at least three terminals for connection to an external circuit. A voltage or current applied to one pair of the transistor's terminals changes the current through another pair of terminals. Because the controlled (output) power can be higher than the controlling (input) power, a transistor can amplify a signal. Today, some transistors are packaged individually, but many more are found embedded in integrated circuits.

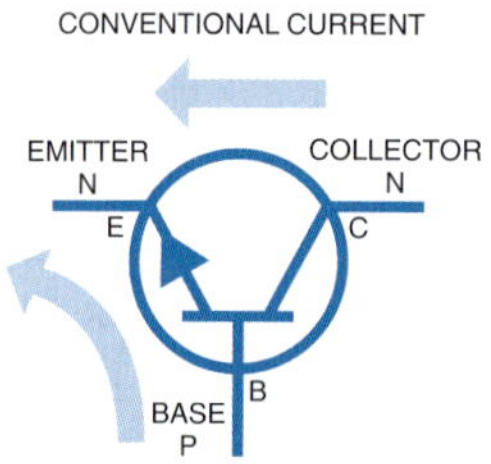

a. Schematic Symbol of NPN Transistor

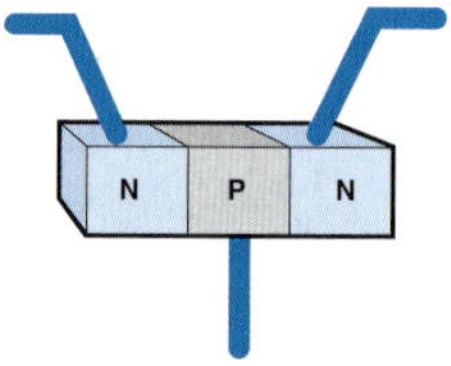

b. Silicon Configuration Suggested by the Symbol

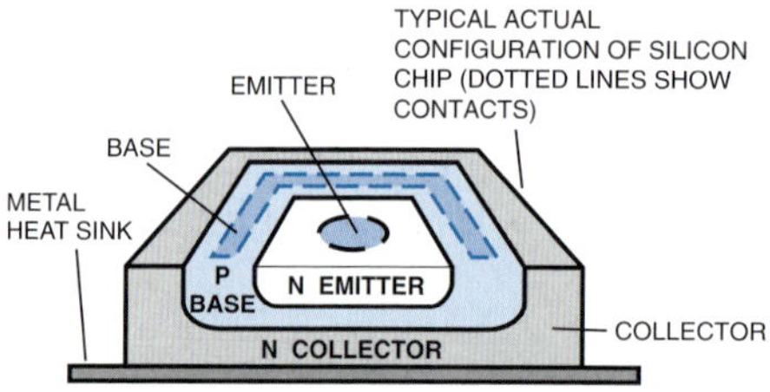

c. Diffused Sandwich Construction

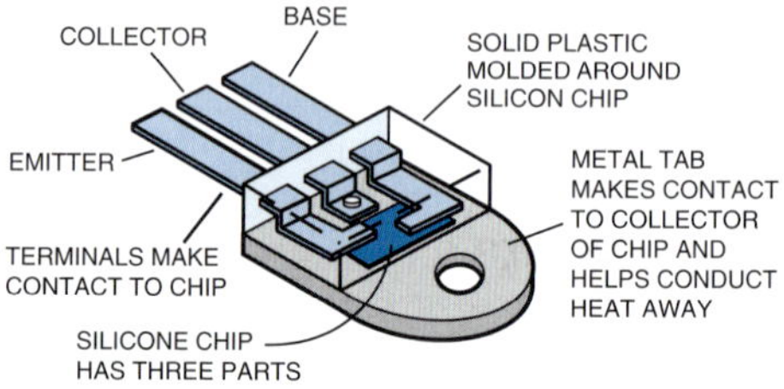

d. Transistor in Package

e. Transistors in an IC package

T6B03 Which of these components can be used as an electronic switch?

A. Varistor
B. Potentiometer
C. Transistor
D. Thermistor

Here is ***the transistor*** – capable of amplification, as well as acting like an on and off switch. **ANSWER C.**

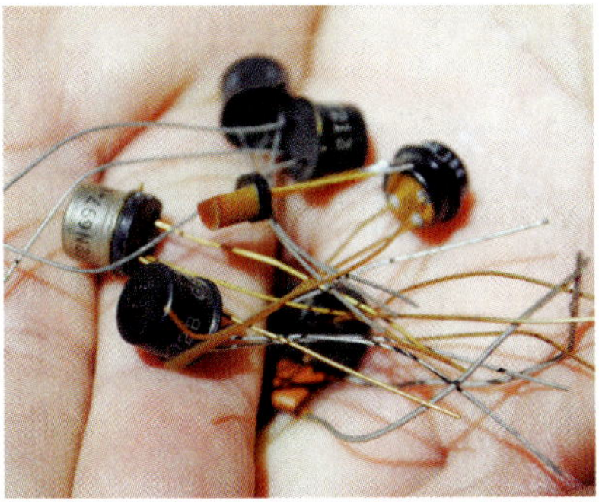

A handful of vintage germanium transistors.

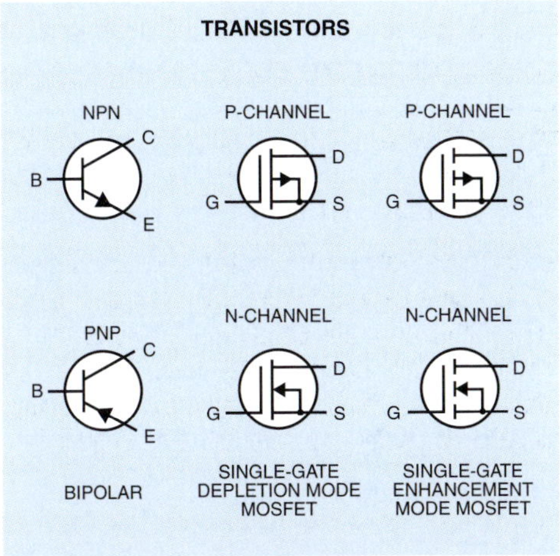

T6B12 What are the names of the electrodes of a bipolar junction transistor?

A. Signal, bias, power
B. Emitter, base, collector
C. Input, output, supply
D. Pole one, pole two, output

Like the FET, the bipolar junction transistor (BJT) has three elements, ***EMITTER, BASE, and COLLECTOR,*** but the term *base* doesn't give you a hint of what it does, like the term *gate* does in an FET. While the operation of a BJT is entirely different than an FET, the FUNCTION of the BJT is about the same: you use one small current to control the flow of a much larger current. **ANSWER B.**

T6B11 What is the term that describes a device's ability to amplify a signal?

A. Gain
B. Forward resistance
C. Forward voltage drop
D. On resistance

We use the word ***GAIN*** to describe how much a device is able to ***amplify a weak signal***. **ANSWER A.**

T6B10 Which of the following can provide power gain?

A. Transformer
B. Transistor
C. Reactor
D. Resistor

Here is the perfect example of NFL (No Free Lunch). You can't get something for nothing, even though it may sometimes seem like it. When we use the term gain, we mean an increase in power. While a simple transformer can give us a step-up in voltage we don't have any increase in power, because the current is proportionally decreased. The only device listed that can actually give us some ***power gain*** is the ***transistor***. It draws more current, hence NFL! **ANSWER B.**

T6B04 Which of the following components can consist of three regions of semiconductor material?

A. Alternator
B. Transistor
C. Triode
D. Pentagrid converter

The transistor is the fundamental solid-state amplifier. The three-layer "bipolar" transistor comes in two flavors, the NPN and the PNP, where N is a negative-doped semiconductor material and P is a positive-doped semiconductor material. Fortunately, you don't need to know a great deal about the internal physics of a ***transistor***, but you do need to understand how the device behaves and is used in radio electronic circuits. **ANSWER B.**

T6B05 What type of transistor has a gate, drain, and source?

A. Varistor
B. Field-effect
C. Tesla-effect
D. Bipolar junction

An ***FET or Field Effect Transistor*** has a source, a gate, and a drain. The gate controls the current from the source to the drain (just like a valve in a piece of plumbing). In operation, the FET is far more like a vacuum tube than a standard *Bipolar* transistor. FETs have extremely high input impedance and are ideal for circuits like active antennas and oscilloscope voltage probes. **ANSWER B.**

T6B08 What does the abbreviation FET stand for?

A. Frequency Emission Transmitter
B. Fast Electron Transistor
C. Free Electron Transmitter
D. Field Effect Transistor

Modern ham radio equipment uses powerful transistors called field effect transistors ***(FET)***. ***The field effect transistor*** can accomplish many functions within our ham radio equipment, providing voltage amplification, where small changes in gate voltage can result in large changes in current flow, leading to voltage amplification. **ANSWER D.**

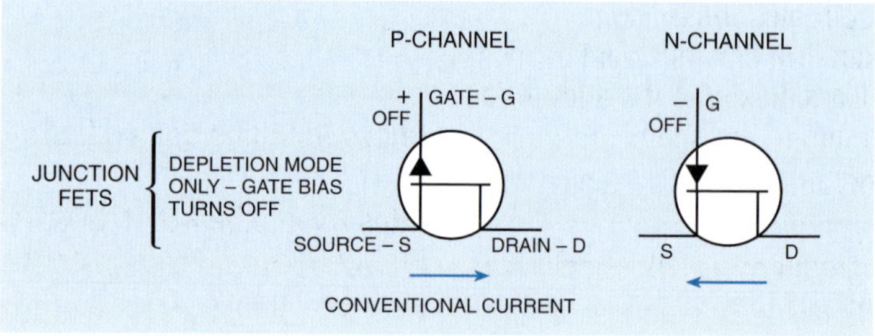

Junction FETs

Website Resources

▼ IF YOU'RE LOOKING FOR	▼ THEN VISIT
Learn More about Electricity	www.forrestmims.com www.masterpublishing.com www.w5yi.org www.arrl.org www.allaboutcircuits.com www.lbl.gov/MicroWorlds/teachers/movingelectrons.pdf
Naval Electrical & Electronics Training System. Oldest, most comprehensive electronics course in existence. Highly recommend for the serious electronics student	www.electriciantraining.tpub.com
Circuit Analysis for Dummies isn't just for dummies! This has a clear and useful description of Kirchhoff's laws, as well as Ohm's Law.	http://www.dummies.com/education/science/science-electronics/three-essential-laws-for-working-with-circuits/
Mr. Carlson's Lab is one of the best video electronics tutorials in cyberspace. He covers a vast array of electronics topics, and it's a lot of fun to watch him in the shop.	https://youtu.be/qqmegXoB7IA
Best general electronics training on the web	https://youtu.be/4OhHTUfKMTw
Another great general electronics site	https://youtu.be/BrairGmZjSM
General Electrical Wiring	https://youtu.be/KnqMJeNN1ug

Ohm's Law & the Magic Circle

Your Granddaddy Ham likely knew that a chap named Georg Simon Ohm (1789-1854) experimented with electricity and discovered that the resistance of a conductor depends on its length in feet, cross-sectional area in circular mils, and its resistivity, which is a parameter that depends on the molecular structure of the conductor and its temperature. Sounds complicated, and it is, but his discovery allows us to calculate some important electrical measurements.

Ohm's Law ($E = I \times R$) states the relationship between voltage, current, and resistance in an electrical circuit. It says that the applied electromotive force (E) in volts, is equal to the circuit current (I) in amperes, times the circuit resistance (R) in ohms. The "magic circle" is an easy way to remember Ohm's Law and understand how to solve for E, I, or R when the other 2 quantities are known. Here is the magic circle and the 3 equations:

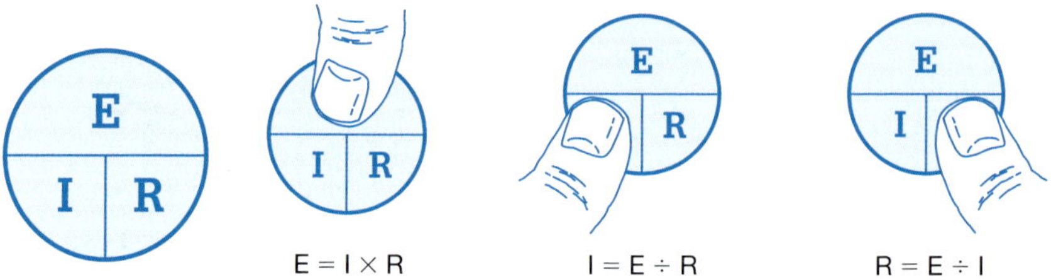

$E = I \times R$
Finding Voltage

$I = E \div R$
Finding Current

$R = E \div I$
Finding Resistance

To use the circle, cover the unknown quantity with your finger and solve the equation using the 2 known quantities. If you know the values of I and R and want to find the value of E, cover the E in the magic circle and it shows that you must multiple I times R. If you want to find I, cover the I and it shows that you must divided E by R. If you want to find R, cover the R and it shows you must divide E by I.

There is another "magic circle" to help you remember how to calculate power in a circuit. Power in watts (P) is equal to current (I) in amperes times volts (E) in volts. Use it the same way as the Ohm's Law magic circle; that is, cover the unknown quantity with your finger and perform the mathematical operation represented by the remaining quantities.

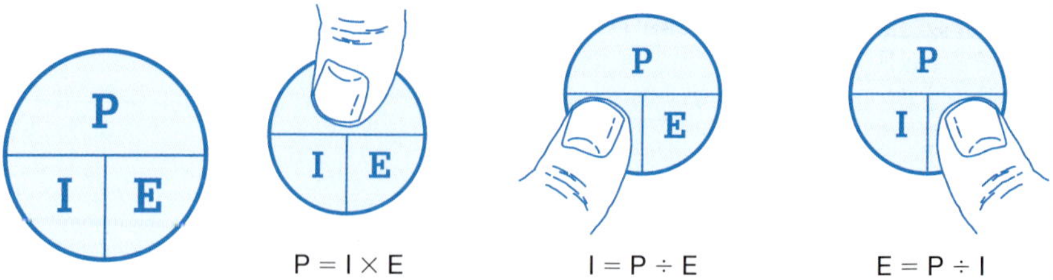

$P = I \times E$
Finding Power

$I = P \div E$
Finding Amperes

$E = P \div I$
Finding Voltage

It's The Law, Per Mr. OHM!

T5A10 Which term describes the rate at which electrical energy is used?

A. Resistance
B. Current
C. Power
D. Voltage

Go outside your home or condo and gaze at the energy meter measuring the rate at which electrical energy is being consumed by all the gadgets inside your new ham shack! That energy meter on the side of your house is called a watt-hour meter, and it is calculating your household electricity usage by calculating ***power*** times time, telling the meter reader how much electricity was used to keep your radio station powered up! Luckily, your modern ham gear won't consume much energy so your power bill won't go up by more than a few dollars per month! **ANSWER C.**

T5A02 Electrical power is measured in which of the following units?

A. Volts
B. Watts
C. Watt-hours
D. Amperes

Power is energy divided by time and is measured in ***watts***. In familiar electrical terms, Power is Voltage times Current (P = I X E). It's as easy as pie! **ANSWER B.**

T5C08 What is the formula used to calculate electrical power (P) in a DC circuit?

A. P = I × E
B. P = E / I
C. P = E – I
D. P = I + E

Power in watts is ***equal to volts times current*** in amps. A 100-watt light bulb, running on 110 VAC house voltage, will draw about 1 amp. The magic circle for power is: P over I E. Cover the unknown quantity with your finger, and perform the mathematical operation represented by the remaining quantities. **ANSWER A.**

T5C09 How much power is delivered by a voltage of 13.8 volts DC and a current of 10 amperes?

A. 138 watts
B. 0.7 watts
C. 23.8 watts
D. 3.8 watts

Power is equal to volts times amps. In this problem, multiple 13.8 volts by 10 amps, and you end up with ***138 watts***. You can do this one in your head. Easy as PIE! **ANSWER A.**

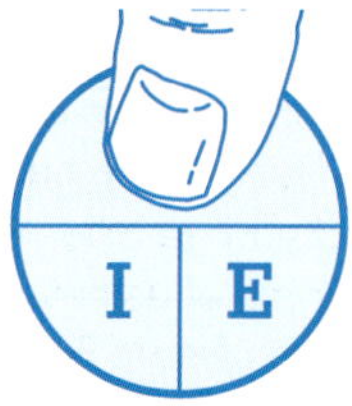

P = I × E
Finding Power

T5C10 How much power is delivered by a voltage of 12 volts DC and a current of 2.5 amperes?

A. 4.8 watts
B. 30 watts
C. 14.5 watts
D. 0.208 watts

Power is equal to volts times amps. Multiple 12 volts by 2.5 amps, and you end up with ***30 watts***. **ANSWER B.**

T5C11 How much current is required to deliver 120 watts at a voltage of 12 volts DC?

A. 0.1 amperes
B. 10 amperes
C. 12 amperes
D. 132 amperes

This time we are calculating for amps, so it is power (120 W) divided by voltage (12 volts DC). Do the keystrokes: Clear ***120 ÷ 12 = 10***. **ANSWER B.**

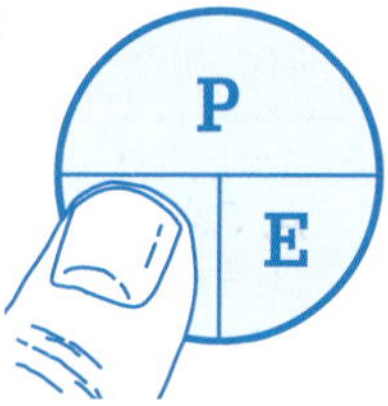

T5D02 What formula is used to calculate voltage in a circuit?

A. E = I x R
B. E = I / R
C. E = I + R
D. E = I – R

Voltage = current x resistance, expressed as E = I x R. This simple formula, called Ohm's Law, states the relationship between voltage, current and resistance in an electrical circuit. **ANSWER A.**

T5D10 What is the voltage across a 2-ohm resistor if a current of 0.5 amperes flows through it?

A. 1 volt
B. 0.25 volts
C. 2.5 volts
D. 1.5 volts

Since we are looking for E in this question, the voltage across a resistor, cover E with your finger, and you now have I (0.5 amps) times R (2 ohms). Simply multiply these two values to obtain your answer of ***1 volt***. On your calculator, which is perfectly legal in the exam room, perform the following keystrokes: Clear 0.5 x 2 = and the answer is 1 volt. Commit the magic circle of success to your memory now! **ANSWER A.**

E = I × R
Finding Voltage

T5D11 What is the voltage across a 10-ohm resistor if a current of 1 ampere flows through it?

A. 1 volt
B. 10 volts
C. 11 volts
D. 9 volts

The question starts out, "What is the voltage across..." so put your finger across E and see that the current in this question is 1 amp through a 10-ohm resistor. One multiplied by 10 is... ***10 volts***. You can do this one in your head. **ANSWER B.**

Check Your Calculator First!

Nothing is more frustrating than a calculator that gives you the wrong answer, especially during a ham radio exam! It's not the calculator's fault when this happens; it's usually because you have violated the order of operation for your particular calculator. Unfortunately, not every calculator uses the same ORDER OF OPERATION. (Trust me; some of us have learned this the hard way!) The time to find this out is *not* in the middle of an exam. Be sure to test your calculator's order of operation before exam time, on a problem with a known answer. You'll be glad you did.

T5D12 What is the voltage across a 10-ohm resistor if a current of 2 amperes flows through it?

A. 8 volts
B. 0.2 volts
C. 12 volts
D. 20 volts

This question is looking for voltage, so we know it's going to be a simple multiplication of 2 amperes through a 10-ohm resistor, with ***20 volts*** as the correct answer. **ANSWER D.**

T5D14 In which type of circuit is voltage the same across all components?

A. Series
B. Parallel
C. Resonant
D. Branch

In a ***parallel circuit***, you have multiple paths for current to flow through. However, the ***voltage is the same across each component***. A circuit with multiple parallel current paths will have more total current than through any single current path. This may seem obvious, but sometimes obvious principles are the most useful! **ANSWER B.**

T5D01 What formula is used to calculate current in a circuit?

A. I = E × R
B. I = E / R
C. I = E + R
D. I = E – R

For ***current***, put your finger over I, and it ***is voltage (E) divided by resistance (R)***. **ANSWER B.**

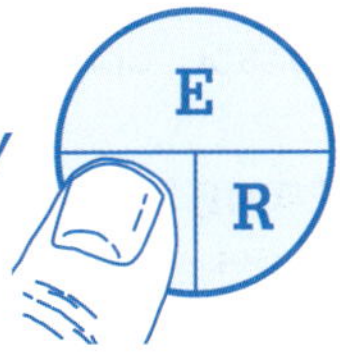

I = E ÷ R
Finding Current

T5D09 What is the current through a 24-ohm resistor connected across 240 volts?

A. 24,000 amperes
B. 0.1 amperes
C. 10 amperes
D. 216 amperes

Do the keystrokes: Clear ***240 ÷ 24 = 10***. Remember, to calculate current, it is voltage on top divided by resistance on the bottom. **ANSWER C.**

T5D07 What is the current in a circuit with an applied voltage of 120 volts and a resistance of 80 ohms?

A. 9600 amperes
B. 200 amperes
C. 0.667 amperes
D. 1.5 amperes

Here they want to know current, so it is voltage (120 volts) divided by resistance (80 ohms). Here are your calculator keystrokes: Clear 120 ÷ 80 = ***1.5***. Be careful that you don't reverse your division — they have incorrect answer C just waiting for you! **ANSWER D.**

KVL & KCL: The Simple Laws of Electrical Plumbing

Two of the most useful principles in electronics are Kirchoff's Voltage Law (KVL) and Current Law (KCL). When you fully understand KVL and KCL every complex electrical circuit becomes a simple plumbing problem, as shown in the simple mechanical drawings.

KVL tells us that the sum of the voltages in a series circuit has to add up to zero. In a series circuit, the sum of the voltage drops across the components has to equal the supply voltage. If you have two resistors in series connected to a battery, the sum of the individual voltage drops across the resistors has to equal the battery voltage.

Each pump creates an output pressure (PSI) that is some value above its input pressure, in this case 20 PSI and 10 PSI respectively. The difference between input and output pressure (pressure differential) is equivalent to the difference of potential across a resistor or a battery. The total accumulated pressure is the sum of the two pumps' pressure differentials (30 PSI), just as the total voltage across two batteries in series is the sum of the individual battery voltages.

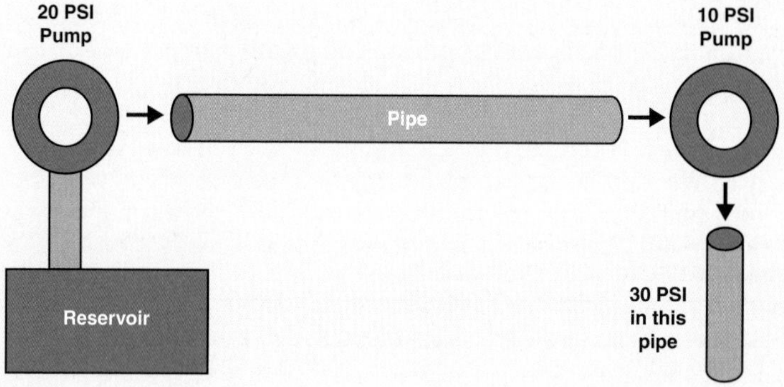

KCL tells us is that the sum of the currents entering a node has to equal the sum of the currents leaving the node (a point of two or more connections). We know that all the water going into one end of a pipe has to come out the other end of the pipe. Now, expand this to the case of a T pipe fitting. We have three ports. If we have 20 gallons per minute going into one port, and 5 gallons per minute flowing out of the second port, we must have 15 gallons per minute flowing out the third port. Replace gallons per minute with amperes and you can solve just about any circuit you will run across.

The current going into the T fitting node (+20 GPM) is designated by the + sign. The currents leaving the node (-5GPM and -15GPM) are designated by the – sign. All three currents have to add up to zero!

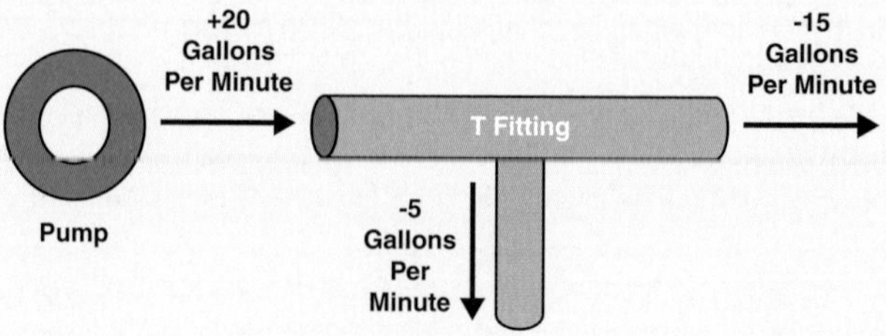

T5D08 What is the current through a 100-ohm resistor connected across 200 volts?

A. 20,000 amperes
B. 0.5 amperes
C. 2 amperes
D. 100 amperes

Be careful on this question – they reversed the order of resistance and voltage that was in the previous question. ***Ohm's Law is rearranged to give: I = E ÷ R. I = 200 Volts ÷ 100 ohms = 2 Amperes.*** **ANSWER C.**

T5D13 In which type of circuit is DC current the same through all components?

A. Series
B. Parallel
C. Resonant
D. Branch

A ***series circuit*** has only one complete path or loop, so the ***same current*** has to go ***through each component***. Also, we know that the sum of the voltage drops across each component in a series circuit adds up to Zero. **ANSWER A.**

T5D03 What formula is used to calculate resistance in a circuit?

A. R = E x I
B. R = E / I
C. R = E + I
D. R = E – I

Put your finger over ***resistance*** in the magic circle and see that it is ***voltage divided by current***. **ANSWER B.**

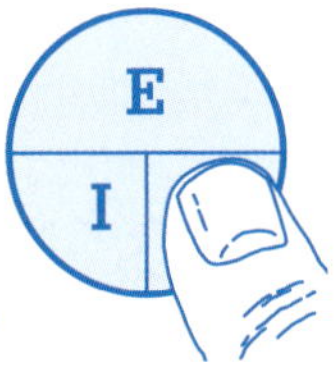

R = E ÷ I
Finding Resistance

T5D04 What is the resistance of a circuit in which a current of 3 amperes flows when connected to 90 volts?

A. 3 ohms
B. 30 ohms
C. 93 ohms
D. 270 ohms

Be careful – they list current first which would go in the bottom of your magic circle, and voltage at the top. Keystrokes: Clear Clear ***90 ÷ 3 = 30***. **ANSWER B.**

T5D05 What is the resistance of a circuit for which the applied voltage is 12 volts and the current flow is 1.5 amperes?

A. 18 ohms
B. 0.125 ohms
C. 8 ohms
D. 13.5 ohms

In this problem they list voltage first, which is 12, on the top divided by 1.5 amps on the bottom. Clear Clear ***12 ÷ 1.5 = 8***. Read each question carefully because they switch around voltage and current, yet your magic circle always says to put voltage on the top and current on the bottom when solving for resistance. **ANSWER C.**

T5D06 What is the resistance of a circuit that draws 4 amperes from a 12-volt source?

A. 3 ohms
B. 16 ohms
C. 48 ohms
D. 8 ohms

Now remember, on most Technician Class questions, you divide the larger number by the smaller number, and presto, you end up with the correct answer. ***12 divided by 4 equals 3***, correct? R = E (12 volts) ÷ I (4 amps). Ohm's Law – simple! **ANSWER A.**

Elmer Point: *This section of our book covering three topics—electronic components, Ohm's law, and schematic drawings—begins your study of electronics! Radio is an electromagnetic phenomena and you need to understand a little bit about how electricity works and how the many devices like resistors, capacitors, transistors, and transformers are used to make it perform the way we want it to.*

Want to learn more about electricity and how electronics work? Here are two books we recommend highly for your self-education! And a couple of reference books we strongly suggest for your ham shack.

Getting Started in Electronics *by Forrest M. Mims III is a true classic. It is used by a wide range of people – from junior-high teachers to the U.S. Army to teach the fundamentals of electricity and electronics.*

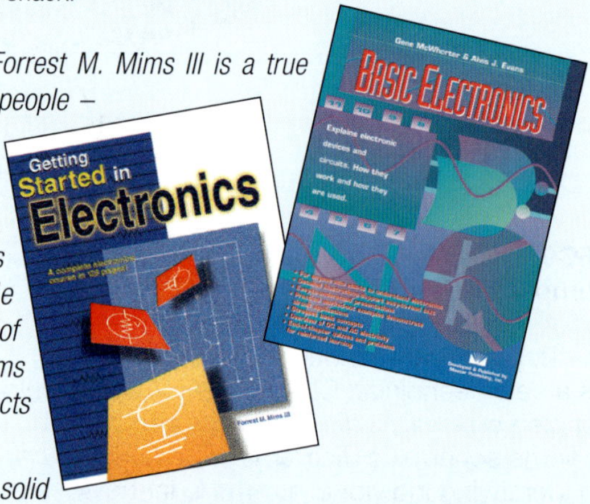

Basic Electronics *by Alvis J. Evans and Gene McWhorter goes a little deeper into the topic, and includes end of chapter quizzes and worked-out problems to teach you in detail the various aspects of electronics.*

Either book – or both – will give you a solid grounding in the theory, science, and practical applications of electronics. You can get you copies of the books, on line at **www.w5yi.org,** *or by calling The W5YI Group at 800-669-9594.*

▼ RECOMMENDED REFERENCE BOOKS FOR YOUR HAM SHACK

ARRL Handbook for Radio Communications with software bundle. This comprehensive reference includes circuit modeling and filter design software, and much more.

RSGB Radio Communications Handbook. Great Britain's answer to the ARRL Handbook provides a more traditional view of Amateur Radio.

Watch This! **https://www.youtube.com/playlist?list=PL1KAjn5rGhizk5f2_whBLJaOHj59OCl_3 (videos 2 + 96)**

Picture This

T6C01 What is the name of an electrical wiring diagram that uses standard component symbols?

A. Bill of materials
B. Connector pinout
C. Schematic
D. Flow chart

As a new Technician Class operator you should be familiar with the identification of simple ***schematic*** diagram illustrations. In the next few questions, we'll take a look at some schematic diagrams with symbols that you should know. We will home in on identifying individual ***symbols*** in the schematic diagrams you'll be seeing on your exam. **ANSWER C.**

T6C12 Which of the following is accurately represented in electrical schematics?

A. Wire lengths
B. Physical appearance of components
C. Component connections
D. All these choices are correct

Schematic diagrams allow us to see exactly ***how components are interconnected***, right down to each and every lead. **ANSWER C.**

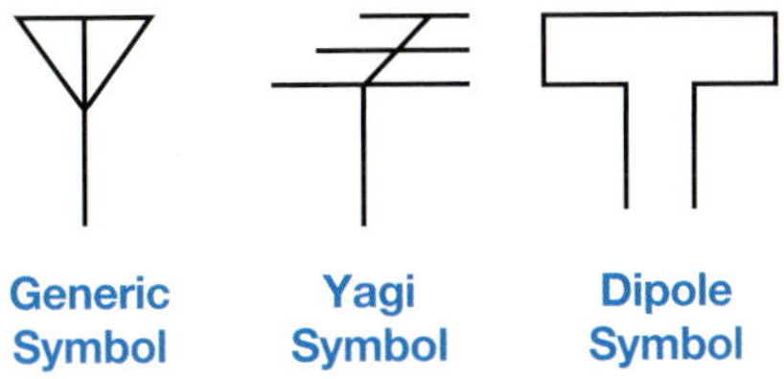

Schematic symbols represent physical electronic components. Some components, like antennas, have a few symbols that represent specific types of components.

T6C10 What is component 3 in figure T-3?

A. Connector
B. Meter
C. Variable capacitor
D. Variable inductor

Component #3 doesn't have a squiggly line like a resistor, but rather a coil type line, so it is a ***variable inductor***. It is variable because we see tap points on the hump lines and a line with an arrow indicating the inductor can be adjusted to any one of the taps. **ANSWER D.**

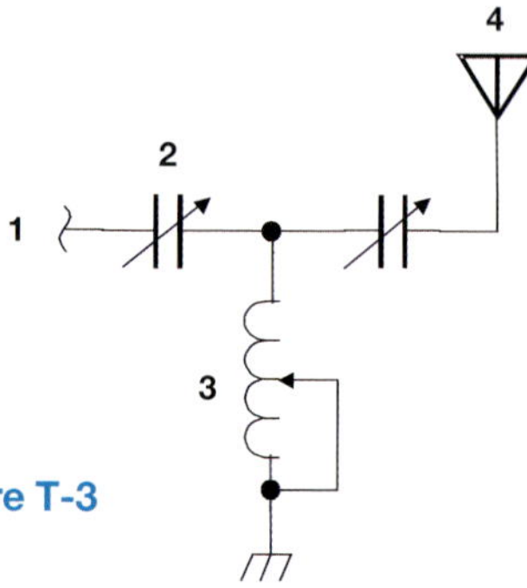

Figure T-3

T6C11 What is component 4 in figure T-3?

A. Antenna
B. Transmitter
C. Dummy load
D. Ground

Component #4 looks like it – an antenna! That ***antenna*** is tuned by some of the preceding circuit components. **ANSWER A.**

T6D08 Which of the following is combined with an inductor to make a resonant circuit?

A. Resistor
B. Zener diode
C. Potentiometer
D. Capacitor

We use series and parallel coils and capacitors to develop a tuned circuit inside your new radio. Another name for a coil is an inductor, and when used with a ***capacitor*** you now have a tuned circuit. **ANSWER D.**

T6D11 Which of the following is a resonant or tuned circuit?

A. An inductor and a capacitor in series or parallel
B. A linear voltage regulator
C. A resistor circuit used for reducing standing wave ratio
D. A circuit designed to provide high-fidelity audio

A ***resonant circuit*** must contain at least one ***inductor*** and one ***capacitor***. It also may have other components. The resonant frequency of the tuned circuit is the frequency at which the inductive reactance and the capacitive reactance are equal. It is one of the most crucial circuits in all of radio. Resonant circuits of one form or another determine the frequency of operation of all radio devices. **ANSWER A.**

T6C02 What is component 1 in figure T-1?

A. Resistor
B. Transistor
C. Battery
D. Connector

This exact figure will be found on your actual Technician Class exam sheet – either with the question or on the last page of your examination materials. Component #1, that squiggly line, is a ***resistor***. Imagine current flowing through all the squiggles creating resistance. **ANSWER A.**

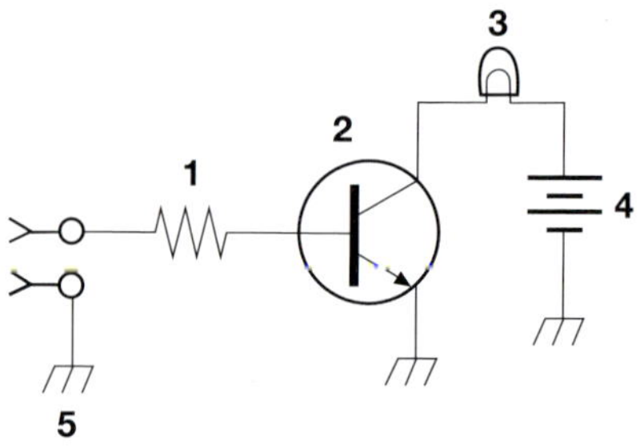

Figure T-1

T6C03 What is component 2 in figure T-1?

A. Resistor
B. Transistor
C. Indicator lamp
D. Connector

Component #2 is our friendly ***transistor***. The arrow is NOT pointing in, so it is an NPN transistor. **ANSWER B.**

T6D10 What is the function of component 2 in figure T-1?

A. Give off light when current flows through it
B. Supply electrical energy
C. Control the flow of current
D. Convert electrical energy into radio waves

Component #2 is a NPN ***transistor***, which ***controls the flow of current***, much like a valve. **ANSWER C.**

T6C04 What is component 3 in figure T-1?

A. Resistor
B. Transistor
C. Lamp
D. Ground symbol

Component #3 looks just like what it is – a small indicator ***lamp***. **ANSWER C.**

T6C05 What is component 4 in figure T-1?

A. Resistor
B. Transistor
C. Ground symbol
D. Battery

It takes voltage to make this circuit work. We get the voltage from the ***battery***, component #4. **ANSWER D.**

T6A12 What type of switch is represented by component 3 in figure T-2?

A. Single-pole single-throw
B. Single-pole double-throw
C. Double-pole single-throw
D. Double-pole double-throw

Your handheld may have one of these – ***a single-pole, single-throw switch***. It is single in both senses because you see only one wire going to the switch and only one single contact point. Single pole-single throw. **ANSWER A**

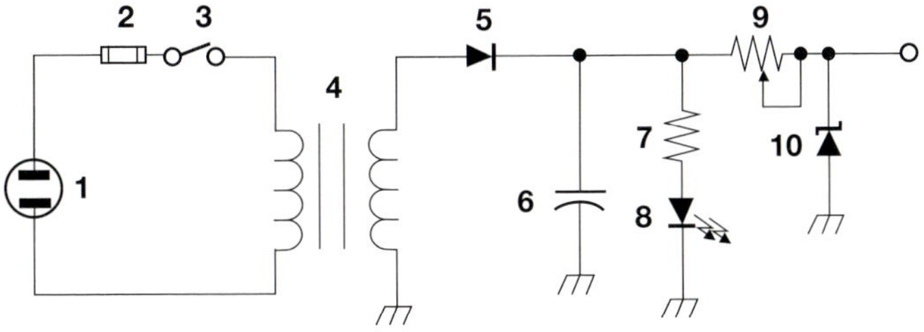

Figure T-2

T6A08 What is the function of an SPDT switch?

A. A single circuit is opened or closed
B. Two circuits are opened or closed
C. A single circuit is switched between one of two other circuits
D. Two circuits are each switched between one of two other circuits

SPDT means Single Pole Double Throw. There is only one *armature,* or movable contact, ***which can make contact with two different stator contacts.*** **ANSWER C.**

T6C09 What is component 4 in figure T-2?

A. Variable inductor
B. Double-pole switch
C. Potentiometer
D. Transformer

Component #4 takes in everything around it and is a ***transformer***. Voltage is passed from the windings on the left to the windings on the right, with the two vertical lines representing an iron core. While a transformer generally increases or decreases voltage based on the number of windings on each side, this transformer looks to have about the same number of turns on the primary and secondary, so the amount of voltage going in will be about the same as the amount of voltage coming out the other side! **ANSWER D.**

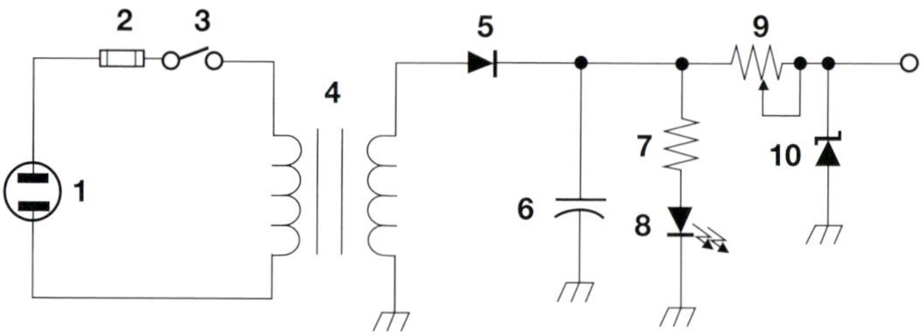

Figure T-2

T6C06 What is component 6 in figure T-2?

A. Resistor
B. Capacitor
C. Regulator IC
D. Transistor

Ok, you're doing so well, let's turn you into a master engineer, and look at figure T2. Component #6 has 2 parallel (sort of) plates, separated by an insulation, so it must be a ***capacitor***. **ANSWER B.**

T6C07 What is component 8 in figure T-2?

A. Resistor
B. Inductor
C. Regulator IC
D. Light emitting diode

See the little arrow symbols on component #8 showing the effects of light? Component # 8 is a ***light emitting diode*** – an LED. Easy! **ANSWER D.**

T6C08 What is component 9 in figure T-2?

A. Variable capacitor
B. Variable inductor
C. Variable resistor
D. Variable transformer

Component #9 is indeed a resistor, but it has a variable tap point on it, so it is a ***variable resistor***. We formally call it a potentiometer. This could be the volume control on your handheld. **ANSWER C.**

T6D04 Which of the following displays an electrical quantity as a numeric value?

A. Potentiometer
B. Transistor
C. Meter
D. Relay

Your larger, high-frequency transceivers may have a mechanical meter to illustrate incoming signal strength. Even if it is an LED or LCD readout, we still call it a signal strength ***meter***. **ANSWER C.**

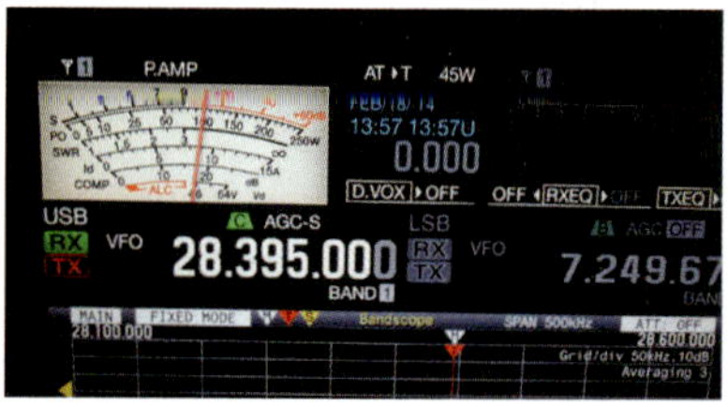

T6D02 What is a relay?

A. An electrically-controlled switch
B. A current controlled amplifier
C. An inverting amplifier
D. A pass transistor

Most of your handhelds don't have one, but a mobile radio that puts out 50 watts will likely contain a relay. The relay is a ***mechanical switch***, opened and closed by current passing through a coil, creating an ***electromagnet***. As soon as the coil is energized, the switch goes from one state to another. **ANSWER A.**

Elmer Point: *Electrical circuits are represented graphically in schematic diagrams using symbols to indicate the various components. Here's a table of schematic symbols that will help you read the diagrams used in this section. If you really get into experimental electronics or kit building, you might want to buy a copy of Forrest Mims's Engineer's Mini Notebook #4,* **Electronic Formulas, Symbols, & Circuits** *available on-line at* **www.w5yi.org** *or by calling the W5YI Group at 800-669-9594.*

SCHEMATIC SYMBOLS

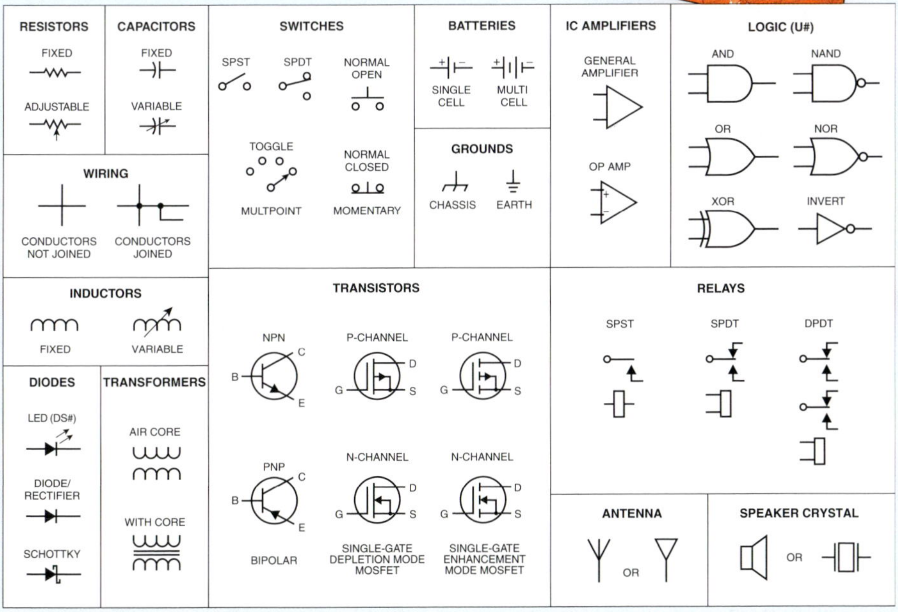

T6D05 What type of circuit controls the amount of voltage from a power supply?

A. Regulator
B. Oscillator
C. Filter
D. Phase inverter

If you want to run your radio on household power, you'll need a power supply that will provide 12 volts for the radio's input circuit. This power supply must have a good ***regulator*** built in so that it does not exceed the 12 volts DC that your radio works with. The AC "wall wart" that came with your handheld is not a regulated power supply: it is a charger. Never transmit from your handheld when plugged in to the AC "wall wart" because there is not enough filtering within that "wall wart" for a good signal and the current draw could damage both the wall wart and your radio. You can turn on the handheld to listen or for programming it while it is connected to the "wall wart", but unplug before transmitting! **ANSWER A.**

Voltage regulator.

T6D06 What component changes 120 VAC power to a lower AC voltage for other uses?

A. Variable capacitor
B. Transformer
C. Transistor
D. Diode

When you purchase your new dual-band handheld it will come with a wall charger that plugs into the side of your radio. The wall charger contains a small ***transformer*** that takes 120 volts AC on the primary and steps it down to a lower AC voltage on the secondary. Diodes and capacitors then filter this lowered AC and convert it to 12 volts DC. The common "wall wart" contains all of this circuitry, and newer "switcher" wall warts are much lighter in weight and the transformer is extremely small in size! **ANSWER B.**

T6D09 What is the name of a device that combines several semiconductors and other components into one package?

A. Transducer
B. Multi-pole relay
C. Integrated circuit
D. Transformer

If you ever look inside the modern ham radio, you'll see rectangular "chips" that are large scale ***integrated circuits***, or "ICs." These chips contain thousands of semiconductors in one nice neat package. **ANSWER C.**

T5B09 Which decibel value most closely represents a power increase from 5 watts to 10 watts?

A. 2 dB
B. 3 dB
C. 5 dB
D. 10 dB

Doubling your power in will lead to an output power increase of 3 dB. Halving your power will lead to a 3 dB decrease. But since we are going from 5 watts to 10 watts, we are doubling our power, and that is a ***3 dB increase***. **ANSWER B.**

T5B10 Which decibel value most closely represents a power decrease from 12 watts to 3 watts?

A. -1 dB
B. -3 dB
C. -6 dB
D. -9 dB

Now we go from 12 watts down to 3 watts by pushing the low power button on our small mobile radio. This is a 4 times decrease (4 x 3 = 12). A 4 times decrease is a power ***decrease of 6 dB***. They are trying to make this one easy for you since doubling results in 3dB change and 4 times is doubling twice so the change is 3 dB twice which is 6 dB. **ANSWER C.**

Decibels

It is important to know about decibels (dB) because they are used extensively in electronics. The decibel is used to describe a change in power levels. It is a measure of the ratio of power output (P_1) to power input (P_2). It is relitavely simple to calculate a dB power change. Remember this table:

dB	Power Change
3 dB	2× Power change
6 dB	4× Power change
9 dB	8× Power change
10 dB	10× Power change
20 dB	100× Power change
30 dB	1000× Power change
40 dB	10,000× Power change

Derivation:

If $\text{dB} = 10 \log_{10} \frac{P_1}{P_2}$

then what power ratio is 20 dB?

$$20 = 10 \log_{10} \frac{P_1}{P_2}$$

$$\frac{20}{10} = \log_{10} \frac{P_1}{P_2}$$

$$2 = \log_{10} \frac{P_1}{P_2}$$

Remember: logarithm of a number is the exponent to which the base must be raised to get the number.

$$\therefore 10^2 = \frac{P_1}{P_2}$$

$$100 = \frac{P_1}{P_2}$$

Or $P_1 = 100\ P_2$

20 dB means P_1 is 100 times P_2

T5B11 Which decibel value represents a power increase from 20 watts to 200 watts?

A. 10 dB
B. 12 dB
C. 18 dB
D. 28 dB

Now we add a linear amplifier. A typical, commercially-built linear amplifier has a gain of about 10 dB (depending on the band and mode). This does not mean you'll be 10 times as loud on the far end if you use a 10 dB amplifier. On HF, a 10 dB difference is about 1-1/2 S-units. Normally, we don't recommend any linear amplifiers as you get started as a Technician Class operator. Going from 20 watts to 200 watts is a bit dangerous, and that is a 10 times increase in power (20 x 10 = 200). ***Ten times increase equals 10 dB***. **ANSWER A.**

T6D07 Which of the following is commonly used as a visual indicator?

A. LED
B. FET
C. Zener diode
D. Bipolar transistor

A good ***visual indicator*** on a handheld radio is the ***LED*** – the light emitting diode that is often used as a transmit indicator. **ANSWER A.**

T6B07 What causes a light-emitting diode (LED) to emit light?

A. Forward current
B. Reverse current
C. Capacitively-coupled RF signal
D. Inductively-coupled RF signal

Another bit of a trick question. Indeed ***a forward current (a positive voltage connected to the anode and a negative voltage applied to the cathode) causes the diode to emit light***. But this by no means explains how it does this. Fortunately, you don't need to know this to answer the question properly. You just need to know you have to plug the LED in the right way to make it work. **ANSWER A.**

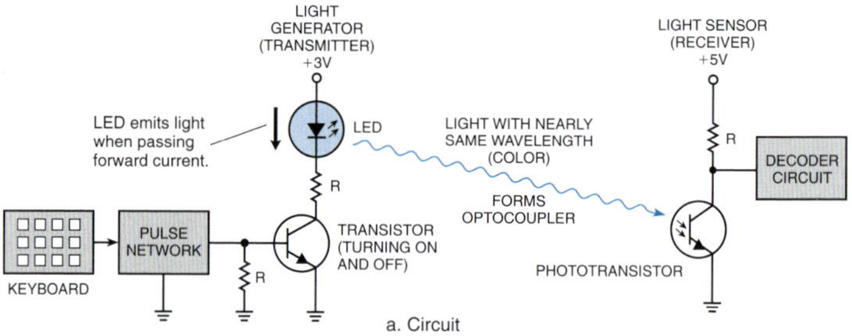

a. Circuit

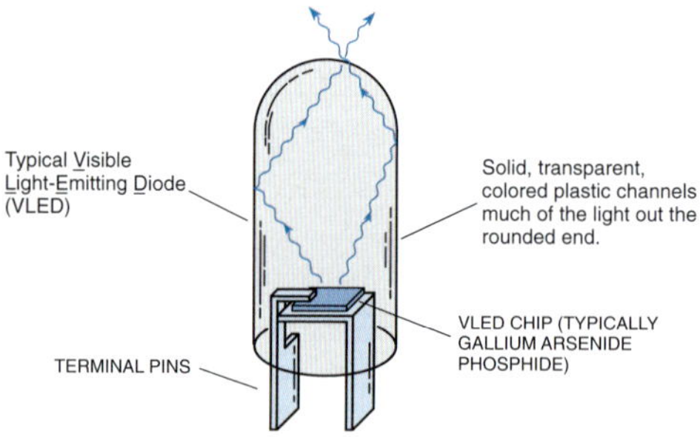

b. Physical Unit for Visible LED

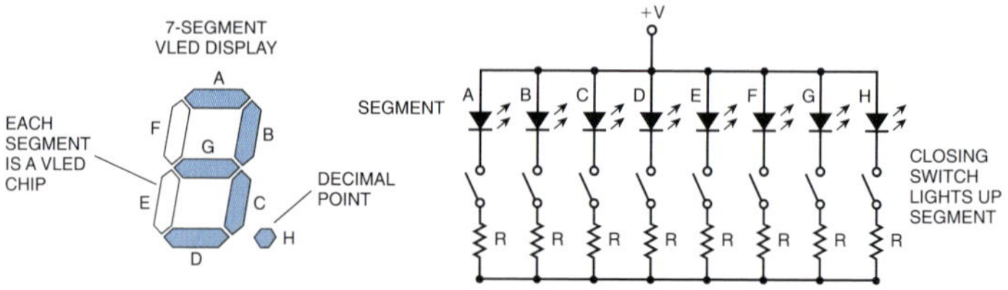

c. VLED Array

T5B02 Which is equal to 1,500,000 hertz?

A. 1500 kHz
B. 1500 MHz
C. 15 GHz
D. 150 kHz

To keep you from running out of pencil lead, we can abbreviate ***1,500,000 Hz*** as either ***1,500 kHz***, or 1.5 MHz. From Hz to kHz, move the decimal 3 places to the left. From kHz to MHz, move it 3 more places to the left. **ANSWER A.**

T5B13 Which is equal to 2425 MHz?

A. 0.002425 GHz
B. 24.25 GHz
C. 2.425 GHz
D. 2425 GHz

There are 1,000 MHz in one GHz, so we simply move the decimal three places to the left (or divide by 1000) to get the final ANSWER, ***2.425 GHz***. **ANSWER C.**

T5B03 Which is equal to one kilovolt?

A. One one-thousandth of a volt
B. One hundred volts
C. One thousand volts
D. One million volts

Remember kilo? ***Kilo means one thousand.*** One kilovolt equals 1,000 volts. Watch out for answer A which is a millivolt: milli for one-thousandth **ANSWER C.**

T5B06 Which is equal to 3000 milliamperes?

A. 0.003 amperes
B. 0.3 amperes
C. 3,000,000 amperes
D. 3 amperes

One milliampere equals one one-thousandth of an ampere (1×10^{-3}); therefore, one ampere equals 1000 milliamperes. Divide milliamperes by 1000 to convert to amperes. Or move the decimal point 3 places to the left.
3000 milliamperes (mA) ÷ 1000 = 3 amperes. **ANSWER D.**

T5B12 Which is equal to 28400 kHz?

A. 28.400 kHz
B. 2.800 MHz
C. 284.00 MHz
D. 28.400 MHz

It's a good idea, and a fun exercise (for some of us, anyway) to memorize all the metric prefixes. They're used a lot in radio electronics. There are 1000 kilohertz in one megahertz, so ***28,400 kHz is equal to 28.400 MHz***. Always remember that if the unit is smaller, you need more of them! Here's another trick. If working with metric prefixes is not second nature to you, always convert the units to the fundamental units while working out a problem. For example, if a question asks you to do calculations involving subtracting XXX kilohertz from YYY megahertz, convert each number in the question to hertz, and then work out the problem. Then convert it back to the unit they want as the last step. This can save you endless confusion, especially in the heat of an exam! Watch out for answer A – it says kHz, not MHz! **ANSWER D.**

Scientific Notation

Prefix	Symbol			Multiplication Factor
exa	E	10^{18}	=	1,000,000,000,000,000,000
peta	P	10^{15}	=	1,000,000,000,000,000
tera	T	10^{12}	=	1,000,000,000,000
giga	G	10^{9}	=	1,000,000,000
mega	M	10^{6}	=	1,000,000
kilo	k	10^{3}	=	1,000
hecto	h	10^{2}	=	100
deca	da	10^{1}	=	10
(unit)		10^{0}	=	1

Prefix	Symbol			Multiplication Factor
deci	d	10^{-1}	=	0.1
centi	c	10^{-2}	=	0.01
milli	m	10^{-3}	=	0.001
micro		10^{-6}	=	0.000001
nano	n	10^{-9}	=	0.000000001
pico	p	10^{-12}	=	0.000000000001
femto	f	10^{-15}	=	0.000000000000001
atto	a	10^{-18}	=	0.000000000000000001

T5B05 Which is equal to 500 milliwatts?

A. 0.02 watts
B. 0.5 watts
C. 5 watts
D. 50 watts

Your handheld transceiver can be dialed down to minimum power output, dramatically conserving battery life. Five hundred milliwatts can be converted to watts by moving the decimal point 3 places to the left. So a handheld at 500 milliwatts output is transmitting ***0.5 watts*** of power, the same as a half watt of power. Believe it or not, you can make many contacts through local repeaters at a half-watt of power, and your batteries will love you for it. **ANSWER B.**

T5B01 How many milliamperes is 1.5 amperes?

A. 15 milliamperes
B. 150 milliamperes
C. 1500 milliamperes
D. 15,000 milliamperes

Your new dual-band handheld might offer as much as 5 to 7 watts of output power. Depending on the battery pack voltage, the transmitter could actually draw as much as 1,500 milliamperes. But don't panic with all those numbers – move the decimal point 3 places to the left to go from milliamps to amps. To convert 1.5 amps to ***1500 milliamperes***, move the decimal point 3 places to the right. **ANSWER C.**

Significant Digits and Decimal Points

Lots of new hams and non-ham electronics technicians tend to go off the rails when it comes to decimal places. Often in my electronics classes, especially when working with AC circuits, some student will ask, "What number do we use for Pi: 3.14159 or 3.1416?" I answer, "How about 3.1?"

If you know about the resistor code, you know that there are only TWO significant digits for any standard resistor (two stripes) plus a multiplier. The same goes for capacitors. Except for special precision components, you will never find any component designated with more than two significant digits. When performing any kind of calculation, there's no point in having 5 or 6 significant digits for Pi (or any other constant) when none of the components you have on hand have better than two significant digits of precision. The accuracy of your calculation is determined only by the LEAST ACCURATE part. Keep in mind that 99% of the radio circuits we used for most of the 20th century were designed with *slide rules*!

2 significant digits of precision is all you will need for any Amateur Radio exam (or even commercial radio FCC exams!) There is a time and place for splitting hairs, but let's keep the significant significant whenever possible.

T5B08 How many microfarads are equal to 1,000,000 picofarads?

A. 0.001 microfarads.
B. 1 microfarad.
C. 1000 microfarads.
D. 1,000,000,000 microfarads.

A picofarad is one millionth (1×10^{-6}) of a microfarad or one million millionth ($1 \times 10^{-6} \times 1 \times 10^{-6} = 1 \times 10^{-12}$) of a farad. Move the decimal point 6 places to the left to convert picofarads to microfarads. You'll find the answer is ***1 microfarad***. **ANSWER B.**

T5B04 Which is equal to one microvolt?

A. One one-millionth of a volt.
B. One million volts.
C. One thousand kilovolts.
D. One one-thousandth of a volt.

We measure the receiver capabilities on a handheld radio in microvolts. The word ***"micro" means one-millionth.*** Mega means million, kilo means thousand, and milli means thousandth. Micro means one one-millionth. **ANSWER A.**

T7D08 Which of the following types of solder should not be used for radio and electronic applications?

A. Acid-core solder.
B. Lead-tin solder.
C. Rosin-core solder.
D. Tin-copper solder.

Acid-core solder must never be used on any electronic circuitry. It will continue to corrode the solder connections creating, at the very least, a poor electrical joint and, at the worst, completely destroy an electronic circuit or printed circuit board (PCB). Never allow any acid core solder in your shop at any time! **ANSWER A.**

If you're going to work on your own electronics gear, "homebrew" ham radio devices, or build kits, learning proper soldering techniques is a fundamental skill.

T7D09 What is the characteristic appearance of a cold tin-lead solder joint?

A. Dark black spots.
B. A bright or shiny surface.
C. A rough or lumpy surface.
D. Excessive solder.

A good solder connection will have a shiny, silvery appearance, and have a very smooth surface. A cold solder joint will have a ***dull gray, grainy and rough surface***. Cold solder connections are caused by too rapid cooling of the connection, or movement of the connection while it is cooling. Never never never blow on a solder connection to "help" it cool down. This is a sure-fire way to create a cold solder joint! **ANSWER C.**

T7D07 Which of the following measurements are made using a multimeter?

A. Signal strength and noise.
B. Impedance and reactance.
C. Voltage and resistance.
D. All these choices are correct.

Every amateur operator should own a ***multimeter***. The multiple function meter can measure ***voltage***, current, and ***resistance***, and check continuity. Even an inexpensive multimeter is better than no meter when you're trying to check out a circuit in the field. You can buy an excellent multimeter for less than $25.00. The important difference between a $25 meter and a $300 meter isn't so much its accuracy, but its ruggedness and its ability to be used with high energy circuits. If you want to see what can go wrong, visit: **http://en-us.fluke.com/training/training-library/electrical-safety/arc-flash-arc-blast/what-is-the-difference-between-arc-flash-and-arc-blast.html**. Please observe the recommendations carefully! Most experienced electronics technicians eventually end up doing the above and more to their meters, so they prefer to get the best meters they can afford! **ANSWER C.**

Parameter	Basic Unit	Measuring Instrument
Voltage (E)	Volts	Voltmeter
Current (I)	Amperes	Ammeter
Resistance (R)	Ohms	Ohmmeter
Power (P)	Watts	Wattmeter

T7D11 Which of the following precautions should be taken when measuring in-circuit resistance with an ohmmeter?

A. Ensure that the applied voltages are correct
B. Ensure that the circuit is not powered
C. Ensure that the circuit is grounded
D. Ensure that the circuit is operating at the correct frequency

Any time you are checking a circuit with an ohmmeter, ***make sure the circuit is not energized!*** If you check any circuit with voltage on it, you will probably toast the ohmmeter for life. **ANSWER B.**

T7D06 Which of the following can damage a multimeter?

A. Attempting to measure resistance using the voltage setting
B. Failing to connect one of the probes to ground
C. Attempting to measure voltage when using the resistance setting
D. Not allowing it to warm up properly

You're likely to ***damage*** your brand-new analog multimeter by ***measuring voltage*** if you accidentally leave it ***in the ohms (resistance) reading setting***. **ANSWER C.**

T7D10 What reading indicates that an ohmmeter is connected across a large, discharged capacitor?

A. Increasing resistance with time
B. Decreasing resistance with time
C. Steady full-scale reading
D. Alternating between open and short circuit

This demonstrates the all-important RC time constant, one of the most vital concepts in electronics. A fully discharged capacitor appears as a dead short to a source of current. ***As the capacitor charges, its effective resistance increases***, decreasing the current flow into it. **ANSWER A.**

Antennas

T9A03 Which of the following describes a simple dipole oriented parallel to Earth's surface?

A. A ground-wave antenna
B. A horizontally polarized antenna
C. A travelling-wave antenna
D. A vertically polarized antenna

Plenty of new Technician Class operators, looking to work some skywaves on 6- and 10-meters, create their own halfwave dipole antennas. The ***dipole antenna*** is usually mounted parallel to the Earth's surface to transmit ***horizontal*** waves. **ANSWER B.**

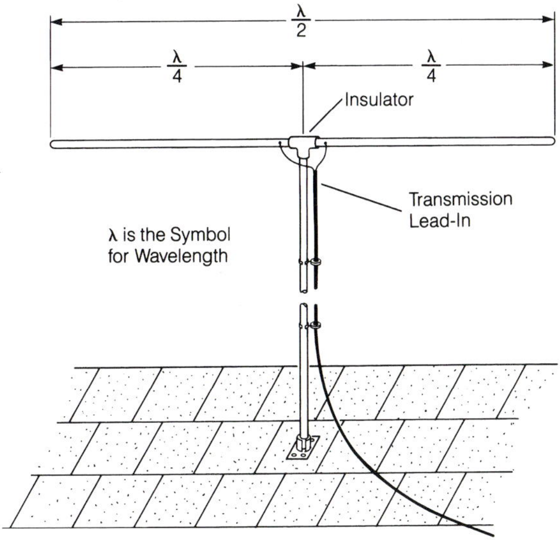

Dipole Antenna

T9A10 In which direction does a half-wave dipole antenna radiate the strongest signal?

A. Equally in all directions
B. Off the ends of the antenna
C. In the direction of the feed line
D. Broadside to the antenna

If the dipole is erected east-to-west, the energy will go out mostly north-to-south, ***broadside to the antenna***. Slightly droop the dipole ends if you wish to send more energy in other directions. **ANSWER D.**

T9A09 What is the approximate length, in inches, of a half-wavelength 6 meter dipole antenna?

A. 6
B. 50
C. 112
D. 236

To calculate the length, end to end, of a ***6-meter, halfwave wire dipole*** antenna, we first need to convert 6 meters to megahertz. Remember that formula, 300 divided by the frequency in megahertz to equal meters? Three hundred divided by 6 equals approximately 50 MHz, and now apply the formula for a halfwave dipole: 468 ÷ frequency in MHz = a halfwave dipole in feet. Calculator: 468 ÷ 50 = 9.36 feet. Now multiply 9.36 X 12 to convert feet to inches, and you get ***112.3 inches*** end to end. Your coax feed line would connect in the center with a homemade choke. You're a regular whiz on that keypad! **ANSWER C.**

T9A05 Which of the following increases the resonant frequency of a dipole antenna?

A. Lengthening it
B. Inserting coils in series with radiating wires
C. Shortening it
D. Adding capacitive loading to the ends of the radiating wires

You can approximate what band a ham is transmitting on by looking at their dipole. The dipole is usually one-half the wavelength length of the meter band of operation. For instance, you can spot a brand new 6-meter Technician Class dipole as approximately 3 meters long, end to end, with coax feeding to it in the center. To raise the antenna's resonant frequency, ***shorten it*** by twisting back a couple of inches of wire at each end and it ***will work at a slightly higher frequency***. **ANSWER C.**

T9A08 What is the approximate length, in inches, of a quarter-wavelength vertical antenna for 146 MHz?

A. 112
B. 50
C. 19
D. 12

We calculate the length, in feet, of a ***quarter-wavelength*** vertical antenna by ***dividing 234 by the antenna's operating frequency*** in megahertz. Let's try the calculator on this one: Clear Clear ***234 ÷ 146 = 1.6 feet***. Since they want the answer in inches, no big deal. To convert feet and fractions of a foot to inches, multiple by 12. Multiply ***1.6 × 12 = 19 inches approximately***. **ANSWER C.**

Ham Hint: *The performance of all ham radio antennas is influenced by their height above ground. A low antenna, only 1/4 wavelength above the ground, sends signals in all directions, including straight up. They call this "cloud warming" and this wasted vertical energy takes away signal strength that you want down close to the horizon. Elevating the antenna to a half wavelength or more above the earth puts more energy into the lower angle of the radiation pattern. This increases both transmit and receive signal strength by minimizing wasted energy going straight up in to the clouds. So get your antennas up high, and ALWAYS STAY WELL CLEAR OF POWER LINES*

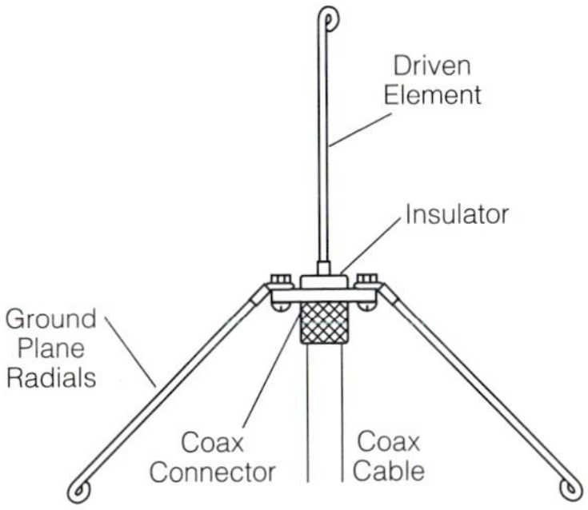

Vertical 1/4 λ Ground-Plane Antenna

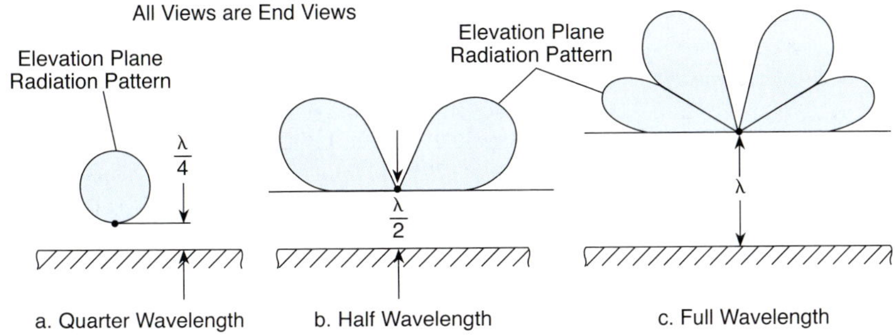

The Radiation Pattern of an Antenna Changes as Height Above Ground is Varied

Source: *Antennas,* A.J. Evans, K.E. Britain, ©1998, Master Publishing, Niles, Illinois

---

T3A05 When using a directional antenna, how might your station be able to communicate with a distant repeater if buildings or obstructions are blocking the direct line of sight path?

A. Change from vertical to horizontal polarization
B. Try to find a path that reflects signals to the repeater
C. Try the long path
D. Increase the antenna SWR

A directional beam antenna for home use may ***bounce a signal*** to and from a distant repeater ***off of a nearby building*** or metal billboard! Trying to perform trick "bank shots" with VHF and UHF antennas is a fun thing to do with surprising results! For decades, hams in Alaska have used Denali as a natural reflector to bounce signals to a good portion of the state! **ANSWER B.**

T9A06 Which of the following types of antenna offers the greatest gain?

A. 5/8 wave vertical
B. Isotropic
C. J pole
D. Yagi

The Yagi antenna, or more properly, the Yagi-Uda antenna, has the most gain possible for a given amount of real estate. It consists of one driven element and one or more parasitic elements, which control the radiation pattern by means of constructive or destructive interference. It is one of the most common steerable beam antennas, and is used extensively on the upper HF amateur bands, well up into the UHF range. **ANSWER D.**

T9A01 What is a beam antenna?

A. An antenna built from aluminum I-beams
B. An omnidirectional antenna invented by Clarence Beam
C. An antenna that concentrates signals in one direction
D. An antenna that reverses the phase of received signals

The beam antenna is much like that new digital television over-the-air antenna on the roof – elements all in line with one another, with the shorter elements in the front. This ***antenna will concentrate signals in one direction***. The beam antenna is subject to the universal principle of NFL – no free lunch. While it concentrates energy in one direction, it subtracts energy in other directions. The total energy radiated by an antenna is always the same, regardless of the gain. The beam antenna is excellent for satellite and weak-signal work, but not necessary for talking around town on 2 meters and 70 centimeters, the 440 MHz band. **ANSWER C.**

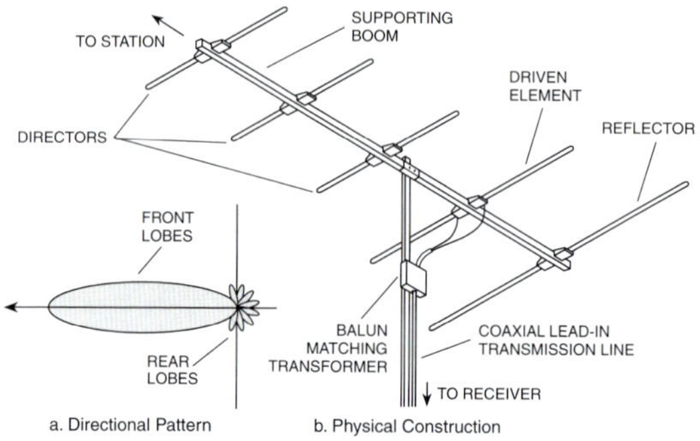

A Beam Antenna — The Yagi Antenna
Source: *Antennas – Selection and Installation*, ©1986, Master Publishing, Inc., Niles, Illinois

Beams are a big attraction for birds! Beam antennas for different bands may be spaced fairly close together on VHF/UHF frequencies.

T8C01 Which of the following methods is used to locate sources of noise interference or jamming?

A. Echolocation
B. Doppler radar
C. Radio direction finding
D. Phase locking

You just added some home office equipment and notice there is a steady, no-voice carrier sitting right on your favorite repeater channel. You know it's coming from your house because your handheld out on the street is just fine. Practice your ***radio direction finding*** skills by going into your home office, tuning in the steady "dead carrier," and start switching off different pieces of equipment and pulling the plug from the AC power receptacle. Remember, some pieces of office equipment will continue to transmit phantom signals, when turned off, yet still plugged in. When you pull the plug that drops the signal, you've found your problem. Time to replace that particular piece of equipment with another brand that hopefully will not cause interference. Worst offenders in the home office? FAX machines, printers, and older telephone modems. **ANSWER C.**

T8C02 Which of these items would be useful for a hidden transmitter hunt?

A. Calibrated SWR meter
B. A directional antenna
C. A calibrated noise bridge
D. All these choices are correct

A very exciting sport is called "fox hunting." A ham will hide a low-power transmitter on the 2-meter band, usually transmitting around 146.565. Expert transmit hunters will use a ***directional antenna*** to home in on the general signal direction. The closer they get to the transmitter the stronger the signal appears on their handheld. When they get relatively close, they pull off the directional antenna, and start "sniffing" around with their handheld with maybe just a paperclip stuck in the antenna socket. **ANSWER B.**

Ham Hint: *Transmitter hunting is not only a fun sport, but a great skill to work with rescue groups tracking down personal locator beacons and ELTs activated in the mountains, or an EPIRB activated out on the ocean. Transmitter hunting is an international sport, and you better be in good shape to do a lot of running when the competition is right behind you homing in on that tiny hidden signal! WARNING: Any time you are around a lot of other transmitter hunters, be sure to wear protective lenses because everyone is swinging those beam antennas all over the place, and you want to be Mr. Safety.*

Radio Direction Finding is a worldwide sport among hams, and kids love the excitement! The sharp ends on each antenna element on the boom should have soft eye-protectors attached before going on a "fox hunt" in the field.

T9A11 What is antenna gain?

A. The additional power that is added to the transmitter power
B. The additional power that is required in the antenna when transmitting on a higher frequency
C. The increase in signal strength in a specified direction compared to a reference antenna
D. The increase in impedance on receive or transmit compared to a reference antenna

The reference antenna with zero gain is called an isotropic radiator. While an isotropic radiator cannot exist in real life, it is an extremely useful measuring stick or ideal. An ***antenna with gain*** is one that takes energy from somewhere within the antenna pattern and ***radiates the signal in the desired direction of transmission***. Got it? It's like taking a reflector to a street lamp and using wasted energy going up and redirecting it with combined energy down at street level. Your mobile VHF and UHF 2 meter and 440 MHz antennas all have a little bit of omni-directional gain, taking energy that would normally go straight up, and radiating it down close toward the horizon. **ANSWER C.**

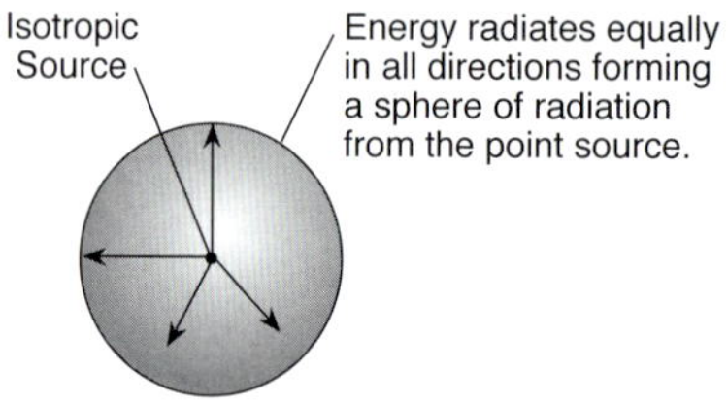

Isotropic Radiator Pattern

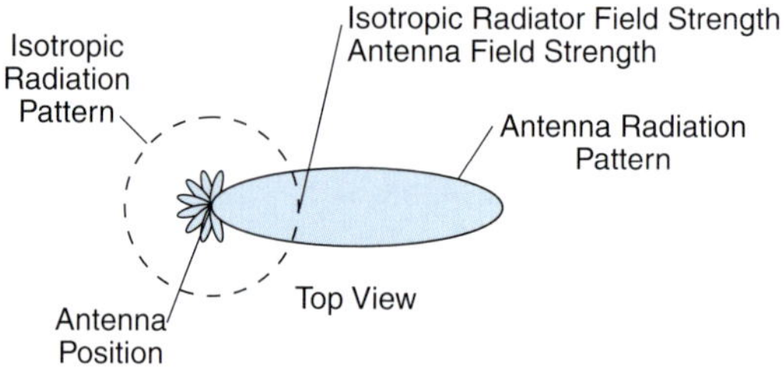

"Gain" of an antenna.

T9A12 What is an advantage of a 5/8 wavelength whip antenna for VHF or UHF mobile service?

A. It has more gain than a 1/4-wavelength antenna
B. It radiates at a very high angle
C. It eliminates distortion caused by reflected signals
D. It has 10 times the power gain of a 1/4 wavelength whip

A ***5/8 wave antenna*** has a more "squashed" pattern than the normal "doughnut" pattern of a quarter wave antenna. It has slightly higher gain than a quarter wave antenna, but it is still omnidirectional. It is not always the best solution however; its ***lower radiation angle*** may cause it to "undershoot" another station at a high altitude. Like any antenna installation, some experimentation is usually needed to find the best arrangement. **ANSWER A.**

T3A03 What antenna polarization is normally used for long-distance CW and SSB contacts on the VHF and UHF bands?

A. Right-hand circular
B. Left-hand circular
C. Horizontal
D. Vertical

The question specifies "VHF and UHF" bands. There is seldom any ionospheric skip this high in frequency. Using ***antennas that are horizontally polarized***, hams all over the country work long range SSB/CW using tropospheric ducting when a high pressure weather cell sits over a big region, refracting SSB/CW VHF/UHF signals from the warm air mass overhead. **ANSWER C.**

T3A04 What happens when antennas at opposite ends of a VHF or UHF line of sight radio link are not using the same polarization?

A. The modulation sidebands might become inverted
B. Received signal strength is reduced
C. Signals have an echo effect
D. Nothing significant will happen

Almost all repeater stations throughout the world use vertical polarization for transmit and receive. This means you must use the same polarization on your handheld antenna, whether it be a tiny rubber duck or a long element flexible whip. Keep your handheld and antenna straight up and down, perpendicular to the Earth, for best reception. If you transmit with the handheld ***antenna horizontal*** to the Earth, your received ***signal*** could be as much as ***100 times weaker*** through that distant repeater. Now you wouldn't want that, would you? **ANSWER B.**

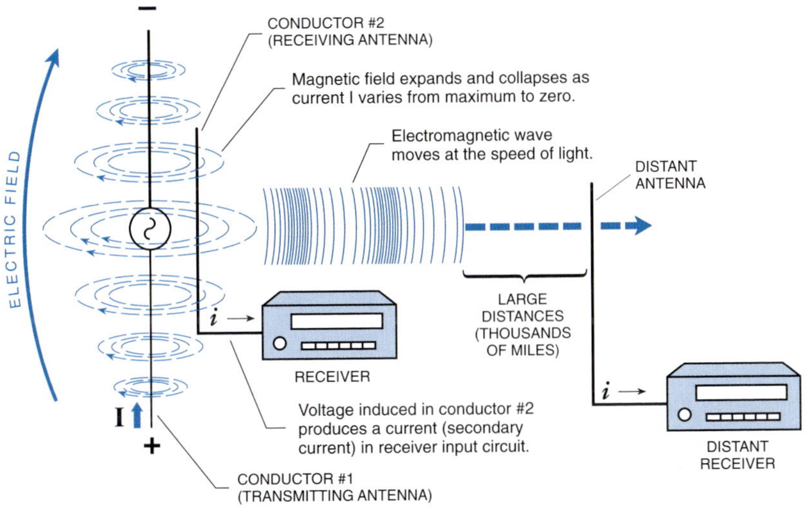

Transmitter to Receiver – Radio waves from transmitting antennas induce signals in receiving antennas as they pass by.

Source: *Listening to Shortwave,* Ken Winters, ©1993 Master Publishing, Inc., Niles, Illinois

T9A02 Which of the following describes a type of antenna loading?

A. Electrically lengthening by inserting inductors in radiating elements
B. Inserting a resistor in the radiating portion of the antenna to make it resonant
C. Installing a spring in the base of a mobile vertical antenna to make it more flexible
D. Strengthening the radiating elements of a beam antenna to better resist wind damage

Antennas are often shorter than they need to be for best performance. ***An antenna that's shorter than its resonant length can be made resonant by adding inductance in series with the antenna***; this inductance is generally known as a loading coil. **ANSWER A.**

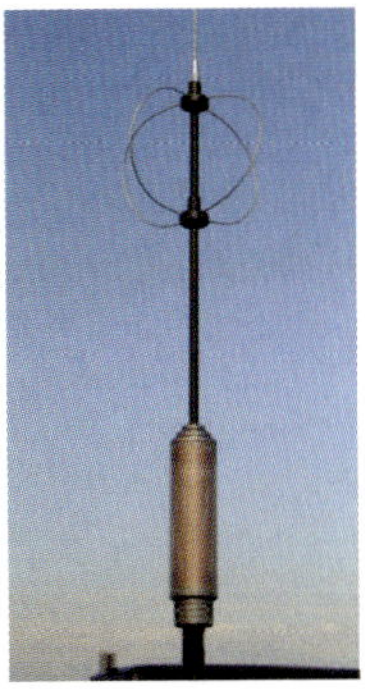

Loading coils like this are used to make an antenna electrically longer.

Website Resources

IF YOU'RE LOOKING FOR	THEN VISIT
Mobile antennas	www.NatCommGroup.com
Satellite antenna	www.ElkAntennas.com
Monster beam antennas	www.Steppir.com

Watch This! **https://www.youtube.com/playlist?list=PL1KAjn5rGhizk5f2_whBLJaOHj59OCl_3 (videos 74, 141, 213, + 250)**

Where To Get Stuff

This year marks my 50th year as a Radio Amateur. I've done just about everything one can do in ham radio... at least once. I've had a vast amount of radio equipment pass through my shack at one time or another, some of it the finest radio equipment ever made anywhere. But in my half century of hamming, I've only had one piece of brand new, store-bought ham radio equipment: my trusty Ten Tec Jupiter. Almost all the rest I've acquired on a "If you can fix it, you can have it" basis. It's a great way to do radio. As I'm fond of saying, "Amateur Radio isn't something you buy; it's something you learn."

Admittedly, a lot of my ham radio gear I acquired as a result of being a broadcast engineer for a few decades; there was always perfectly good but "well-seasoned" equipment radio stations were trying to get rid of. However, it was Amateur Radio that gave me the skills to become a broadcast engineer in the first place. I would never have acquired those skills if I had just bought a bunch of radios.

At every ham radio flea market or hamfest there are *tons* (sometimes quite literally!) of perfectly serviceable amateur radio gear that can be picked up for next to nothing and brought up to like-new condition with just a little TLC. This is especially true of HF equipment. We always hear new hams complain about the price of HF equipment, yet there are thousands of spare HF rigs in shacks all across the country. Most Elmers are happy to part with these transceivers for a mere to any young ham who shows an interest.

Here's a little secret. If you look through the ads in any of the ham radio publications, the manufacturers usually show their top-of-the-line "flagship" models. But every major manufacturer has lower-tier, moderately priced rigs just right for new hams only wanting to get their feet wet. So stay away from the glossy photo section of the ads. There may not be a photo at all of some of the "starter" rigs that you'll want.

Don't confine your search to the major "The Big 3" manufacturers. There are dozens of small "cottage" manufacturers of simple rigs and kits. You may need to search a little to find them as they don't always advertise in the ham magazines. QRP Labs **https://www.qrp-labs.com** is one shining example: their QRP rigs are the simplest and cheapest way I know of to get on HF, and they are really magnificent little radios.

Finally, there is one piece of radio equipment that *every* radio amateur needs, regardless of license class: a general coverage shortwave receiver of *any* quality. This is the *best* way, by far, to learn how radio works; generations of hams first learned about radio this way. Propagation is getting better all the time, and even the most inexpensive shortwave receiver will give you an open window to the world.

Feed Me With Some Good Coax

T9B03 Why is coaxial cable the most common feed line for amateur radio antenna systems?

A. It is easy to use and requires few special installation considerations
B. It has less loss than any other type of feed line
C. It can handle more power than any other type of feed line
D. It is less expensive than any other type of feed line

When you get started as a new Technician Class operator, most of your external antenna work will almost always include a feed line called ***coaxial cable***. Like a garden hose, don't kink it, scrunch it, chop it, or squash it in a car door. As long as you are careful to keep coax cable in its natural round shape, it ***requires few installation considerations*** and can be run right alongside the vehicle frame.
ANSWER A.

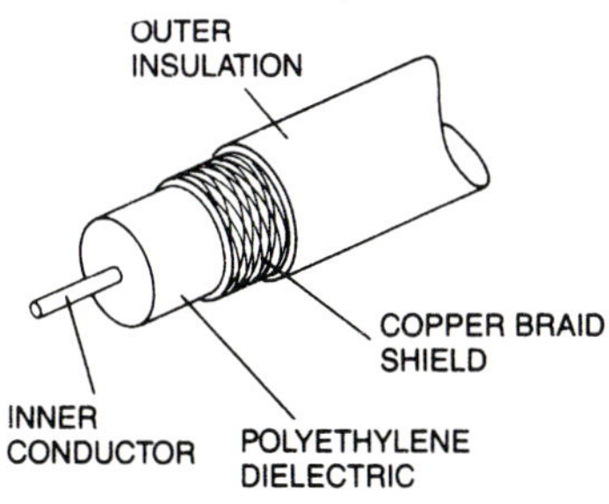

Coaxial (Called Coax)

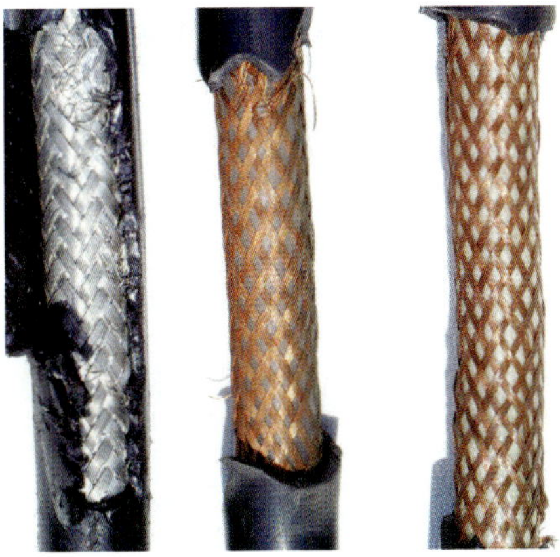

T9B02 What is the most common impedance of coaxial cables used in amateur radio?

A. 8 ohms
B. 50 ohms
C. 600 ohms
D. 12 ohms

Most ***ham radio coax is rated at 50 ohms*** impedance. But if you accidentally slam a car door on it and distort its round shape, now the impedance will be more like 20 ohms, which will result in a mismatch and reduced performance. **ANSWER B.**

Coax choke.

T9B05 What happens as the frequency of a signal in coaxial cable is increased?

A. The characteristic impedance decreases
B. The loss decreases
C. The characteristic impedance increases
D. The loss increases

The higher we go in frequency on 2 meters, 440 MHz, 900 MHz, and 1270 MHz, the greater the need for larger diameter coaxial cable. The ***higher*** we go in ***frequency***, the ***greater the loss*** of energy within the coax cable. While you can get by with pencil thin round shaped coax for HF frequencies, always try to use larger cable, about the size of your thumb, for most ham radio VHF and UHF antenna installations. **ANSWER D.**

T9B07 Which of the following is true of PL-259 type coax connectors?

A. They are preferred for microwave operation
B. They are watertight
C. They are commonly used at HF and VHF frequencies
D. They are a bayonet-type connector

The common ***PL-259 connector*** is what we normally use on ***high frequency radios*** and is common on your little mobile and base 2 meter/440 MHz radio, too. **ANSWER C.**

From left to right: BNC, PL-259, Type N, SMA, and cable television F connector.

T9B06 Which of the following RF connector types is most suitable for frequencies above 400 MHz?

A. UHF (PL-259/SO-239)
B. Type N
C. RS-213
D. DB-25

If you plan to operate ***above 400 MHz***, such as working satellites on 70 cm, or at 1270 MHz, the antennas will usually accept only a ***type N connector***. If you try to screw in a common PL-259 connector to a type N receiver, you will permanently damage the antenna connection. Look carefully to make sure you know the difference between the type N connector and the more common PL-259. Always use Type N above 400 MHz! **ANSWER B.**

T7C11 What is a disadvantage of air core coaxial cable when compared to foam or solid dielectric types?

A. It has more loss per foot
B. It cannot be used for VHF or UHF antennas
C. It requires special techniques to prevent moisture in the cable
D. It cannot be used at below freezing temperatures

Your buddy down the street has a very special gift for you – air core coaxial cable. Tell him thanks, but no thanks. This is professional cable for commercial communications and requires special connectors plus dry nitrogen gas feeding down the inside, and it's like working with a frozen garden hose. Stick with regular ham radio quality coaxial cable and you'll be set! That ***air core*** line is good but ***requires special techniques to keep the water out***. **ANSWER C.**

Large coax, with hollow center conductor, low loss.

T7C09 Which of the following causes failure of coaxial cables?

A. Moisture contamination
B. Solder flux contamination
C. Rapid fluctuation in transmitter output power
D. Operation at 100% duty cycle for an extended period

Moisture getting into coax cable is the number one cause of a lousy signal on the air. Always seal exposed coaxial cable connectors at the antenna feed point. Regular coax PL-259 connectors are not water-tight. **ANSWER A.**

T9B08 Which of the following is a source of loss in coaxial feed line?

A. Water intrusion into coaxial connectors
B. High SWR
C. Multiple connectors in the line
D. All these choices are correct

While all coaxial transmission lines have *some* degree of loss, mainly due to copper resistance ("copper loss"), the losses can increase due to a number of factors: ***water intrusion into the connectors, high SWR, or multiple lossy connectors*** in the line. **ANSWER D.**

T7C10 Why should the outer jacket of coaxial cable be resistant to ultraviolet light?

A. Ultraviolet resistant jackets prevent harmonic radiation
B. Ultraviolet light can increase losses in the cable's jacket
C. Ultraviolet and RF signals can mix, causing interference
D. Ultraviolet light can damage the jacket and allow water to enter the cable

Always buy professional-grade coaxial cable. Quality coax has an outer jacket that will not break down in ultraviolet light from the Sun. If you try to use old CB radio coax, the ***jacket*** may already be ***breaking down, allowing moisture to enter the cable***. You'll be off the air in an instant when the next rain hits! **ANSWER D.**

T9B10 What is the electrical difference between RG-58 and RG-213 coaxial cable?

A. There is no significant difference between the two types
B. RG-58 cable has two shields
C. RG-213 cable has less loss at a given frequency
D. RG-58 cable can handle higher power levels

RG-58 coaxial cable is very small and losses are high. ***RG-213 coax*** is the "big stuff" with ***less loss*** at any frequency. We call it LMR-400 "style" coax as there are many similar sizes of this coax with even better internal construction to minimize losses. **ANSWER C.**

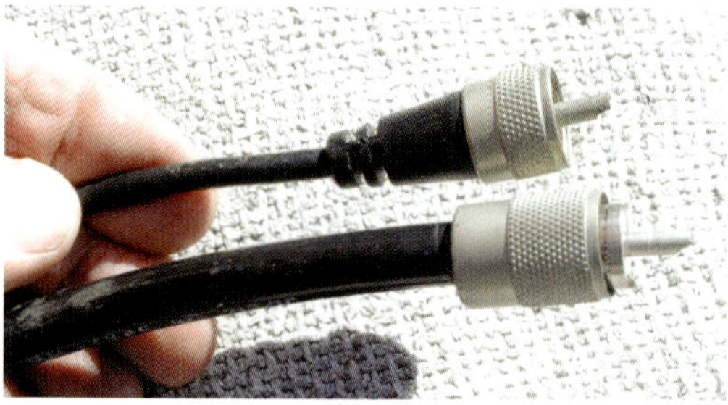

Larger coax, lower losses.

Coax Cable Type, Size and Loss per 100 Feet

Coax Type	Size	Loss at HF 100 MHz	Loss at UHF 400 MHz
RG-58U	Small	4.3 dB	9.4 dB
RG-8X	Medium	3.7 dB	8.0 dB
RG-8U	Large	1.9 dB	4.1 dB
RG-213	Large	1.9 dB	4.5 dB
Hardline	Large, Rigid	0.5 dB	1.5 dB

T9B11 Which of the following types of feed line has the lowest loss at VHF and UHF?

A. 50-ohm flexible coax
B. Multi-conductor unbalanced cable
C. Air-insulated hardline
D. 75-ohm flexible coax

If you're getting real serious about operating with satellites you may wish to locate that guy down the street who had a spool of ***air-insulated hardline coax***. Ask him for a hunk of it and let him do the connections, too. Air-insulated hardline is the ultra-ultra best for weak signal work on VHF and UHF but requires a big deal in getting all the connectors properly soldered and each end sealed against the elements with non-conducting goo called Coax Seal™. **ANSWER C.**

T7C02 Which of the following is used to determine if an antenna is resonant at the desired operating frequency?

A. A VTVM
B. An antenna analyzer
C. A Q meter
D. A frequency counter

Most 2 meter/440 MHz mobile whips and white fiberglass base antennas are pre-set to the bands of operation. Down on the worldwide bands, like your voice privileges on 10 meters, you may need to do some antenna adjustments. A handy piece of test equipment is the ***antenna analyzer***. This allows you to work on the antenna safely, on the roof, and then test its resonant frequency without needing to climb down the ladder to transmit on your big ham rig. The little antenna analyzer hooks directly to the base of the antenna and allows you to do the test easily while on the roof. **ANSWER B.**

This SWR analyzer is an important part of your test equipment if you regularly work on tuning antennas.

T9B12 What is standing wave ratio (SWR)?

A. A measure of how well a load is matched to a transmission line
B. The ratio of amplifier power output to input
C. The transmitter efficiency ratio
D. An indication of the quality of your station's ground connection

If you decide to home brew your own antenna system, you'll have fun! But you'll need to obtain some additional test equipment, like a standing wave ratio (SWR) meter. This will allow you to test ***how well*** your antenna, which we call ***the load, is matched to the transmission line impedance*** as measured in your coax feed line. We want both to be 50 ohms for a perfect 1:1 impedance match. **ANSWER A.**

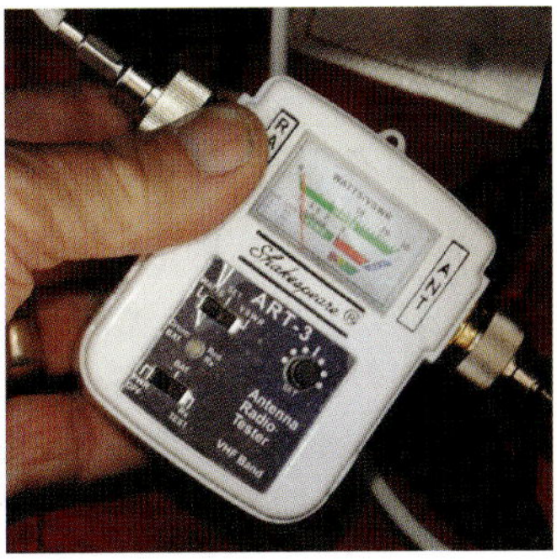

Use a small SWR meter like this one to perform a quick check of your mobile antenna system.

T4A02 Which of the following should be considered when selecting an accessory SWR meter?

A. The frequency and power level at which the measurements will be made
B. The distance that the meter will be located from the antenna
C. The types of modulation being used at the station
D. All these choices are correct

The typical "monimatch" type SWR meter has a sensitivity which increases directly with frequency. If you are interested in *only* the SWR, this is not too much of a problem because the forward and reflected sensitivity increase at the same rate. However, if you are using the meter for absolute power measurement, the meter must be ***calibrated for the power range and frequency***. This becomes more critical at VHF and UHF, and it is advised to get a meter actually designed for these frequencies. **ANSWER A.**

T4A05 Where should an RF power meter be installed?

A. In the feed line, between the transmitter and antenna
B. At the power supply output
C. In parallel with the push-to-talk line and the antenna
D. In the power supply cable, as close as possible to the radio

In most situations, a combination ***SWR and Power Meter*** is most useful immediately ***after the transmitter***, but ***before any antenna*** matching or tuning devices. The antenna tuner (if any), the transmission line, and the antenna form an antenna system, and the SWR meter is a good way to determine that the system is tuned and matched properly to the transmitter. However, there's a lot more to antenna performance than SWR itself, and you should avoid "obsessing" about SWR as is common practice among newcomers and some "old timers" as well. **ANSWER A.**

T9B01 What is a benefit of low SWR?

A. Reduced television interference
B. Reduced signal loss
C. Less antenna wear
D. All these choices are correct

If your antenna match is perfect, you will have a low standing wave ratio and there will be an ***efficient transfer of power*** up the coax to your antenna system, ***with minimum signal losses***. **ANSWER B.**

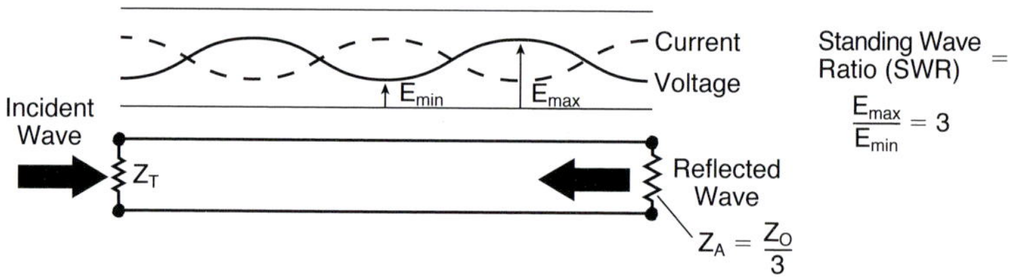

Impedance Mismatch Causes Reflected Wave

T7C04 What reading on an SWR meter indicates a perfect impedance match between the antenna and the feed line?

A. 50:50
B. Zero
C. 1:1
D. Full Scale

The ***perfect match***, as read on a SWR meter, would be ***1 to 1***. **ANSWER C.**

*SWR Reading	Antenna Condition
1:1	Perfectly Matched
1.5:1	Good Match
2:1	Fair Match
3:1	Poor Match
4:1	Something Definitely Wrong

*Constant Frequency

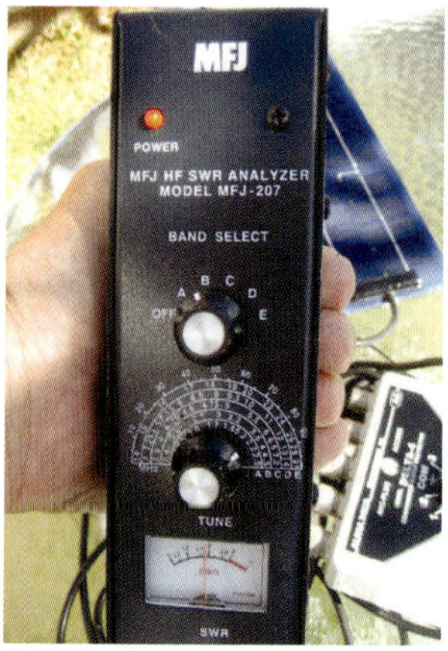

A battery operated SWR analyzer for tower antenna work. To get a better understanding of SWR, watch this excellent video on YouTube: youtu.be/DovunOxlY1k

T7C06 What does an SWR reading of 4:1 indicate?

A. Loss of -4 dB
B. Good impedance match
C. Gain of +4 dB
D. Impedance mismatch

In a typical amateur radio installation, a ***4:1 SWR*** reading is an indication of an ***impedance mismatch*** between your transmission line and your load (your antenna). With that much of a mismatch, all the available power is not being delivered to the antenna. However, where you measure SWR is important. Many effective antenna systems have high SWR readings on the ANTENNA side of a tuner, for instance. However, the SWR that your transmitter "sees" should be as low as possible, ideally 1:1. **ANSWER D.**

T7C05 Why do most solid-state transmitters reduce output power as SWR increases beyond a certain level?

A. To protect the output amplifier transistors
B. To comply with FCC rules on spectral purity
C. Because power supplies cannot supply enough current at high SWR
D. To lower the SWR on the transmission line

A standing wave on a transmission line can cause the impedance that the transmitter "sees" to be either too high or too low in comparison with the nominal 50-ohm impedance the transmitter is designed for. If the impedance is too high, the output transistors will be subject to excessive voltage that could cause them to break down. If the impedance is too low, the current will be excessive, which will cause overheating. ***SWR "foldback" protection reduces the effects of both fault conditions on the transistors*** by reducing the drive level when the SWR is higher than normal. **ANSWER A.**

Most HF rigs have the SWR meter built in, but you can always use an external SWR bridge to double-check your new dipole antenna system.

T9B09 What can cause erratic changes in SWR?

A. Local thunderstorm
B. Loose connection in the antenna or feed line
C. Over-modulation
D. Overload from a strong local station

It's windy today and you notice that your local repeater jumps up and down in signal strength. In fact, you tune around and notice that many other stations are doing the same thing on your external chimney-mounted, dual-band antenna. Guess what? You probably have a ***loose connection up at the antenna or*** a loose coax cable ***feed line connector***. Don't transmit – safely get up on the roof and see what is loose. **ANSWER B.**

Make sure all coax connections are tight to help minimize interference.

Tube Be or Not Tube Be *Working with Hybrid HF Radios*

Your first car was not a Maserati, it was probably a Yugo. Likewise, your first HF radio will very likely not be a top-of-the-line model like those you see in the glossy ham radio magazines. This is very good news if you are just getting started in HF ham radio!

Some of the finest radios ever built are hybrid rigs, meaning they use solid state technology for the receiving section and most of the low power transmitting section, relying on vacuum tubes for the "heavy lifting" RF output stages.

There are probably countless unused HF rigs sitting on shelves in the shacks of local hams around your community. This is true even in a remote location like North Pole, Alaska! It's almost impossible to even give away perfectly serviceable HF equipment to new hams; rigs which many "properly seasoned" hams would give their eye teeth for. The reason for this is basically that some older equipment seems to be more difficult to use. Not so. In fact, a lot of older equipment is far more tolerant and forgiving of operator error than "newfangled" equipment. A large reason for this is the nearly universal use of vacuum tube final RF amplifiers in a lot of older transmitters and transceivers.

Operating a tube radio is not difficult at all. There are only two or three knobs that might be unfamiliar to you, usually a Plate TUNE, a LOAD, and sometimes a PRESELECTOR, which tunes both the receiver input and transmitter DRIVER stage. There is a nearly universal process of tuning (or "dipping") and loading for proper output power regardless of the particular rig you're using. We won't go into the detailed procedure here, as it is much easier to learn from the Elmer from whom you're acquiring the rig! You can learn the procedure in about a minute, and when you have it down pat you can do it in about 15 seconds, with no danger whatsoever to your radio or yourself.

Never be reluctant to pick up an older radio; you'll learn a lot more about radio theory than you are likely to learn with a fully "automatic" radio. Remember, amateur radio is something you learn, not something you buy!

T7C08 Which instrument can be used to determine SWR?

A. Voltmeter
B. Ohmmeter
C. Iambic pentameter
D. Directional wattmeter

The technical ham may substitute an in-line ***directional watt meter*** for an SWR meter. The most popular is manufactured by Bird and contains a rotating sensor element that allows the technical ham to compute power forward and reverse power coming back due to an improperly constructed antenna or a bad feed line. The SWR can be calculated using these two figures. **ANSWER D.**

T7C07 What happens to power lost in a feed line?

A. It increases the SWR
B. It is radiated as harmonics
C. It is converted into heat
D. It distorts the signal

In physics, all power must be used for something. If it is not radiated out the antenna to be heard by others around the world, ***it is converted into heat*** and lost within the system. Think of it as frustrated electrons. They are mad (hot) because they can't do what they were intended for. **ANSWER C.**

T9B04 What is the major function of an antenna tuner (antenna coupler)?

A. It matches the antenna system impedance to the transceiver's output impedance
B. It helps a receiver automatically tune in weak stations
C. It allows an antenna to be used on both transmit and receive
D. It automatically selects the proper antenna for the frequency band being used

An ***antenna tuner will match*** your 6- and 10-meter ***transceiver to an antenna system*** that might not be perfectly tuned to the frequency on which you wish to operate. Antenna tuners are not normally found in 2-meter/440 MHz systems. **ANSWER A.**

Feed Me with some Ladder Line too!

Many new hams (and some older ones too) don't realize that coaxial cable is actually somewhat of a "Johnny-Come-Lately" in the radio world. For the first several decades of amateur radio's existence, coax was never even heard of. It only became readily available to radio amateurs after World War II. So, what did hams use before coax? The transmission line of choice was open-wire feed line, also known as ladder line. It is simply two parallel wires spaced at periodic intervals by insulators. While not as convenient to install as coaxial cable, ladder line has much lower loss than coax, even under conditions of extremely high SWR. (In fact, when ladder line was king, nobody knew what SWR was, nor did the average ham have any way of measuring it!) It's a wonder hams were able to communicate at all with such low tech, and lack of test equipment, but they did, very effectively!

You are not too likely to see true ladder line on the VHF frequencies, but you will encounter a variation of it known as twin lead. Twin lead was used extensively on television and FM receiving antennas and is still incorporated in a lot of homebrew VHF amateur radio antennas. Where ladder line still reigns supreme is on HF, especially the high powered stuff.

T7C01 What is the primary purpose of a dummy load?

A. To prevent transmitting signals over the air when making tests
B. To prevent over-modulation of a transmitter
C. To improve the efficiency of an antenna
D. To improve the signal-to-noise ratio of a receiver

Test your brand-new radios with a ***dummy load***, which ***prevents the signal from being sent out*** on the airwaves. The dummy load lets you look at your signal on other equipment to make sure it's absolutely clean before hooking up to an outside antenna. [Note: In some cases a dummy load is incorporated in the "front end" of a spectrum analyzer, usually limited to a fraction of a watt or so.] **ANSWER A.**

T7C03 What does a dummy load consist of?

A. A high-gain amplifier and a TR switch
B. A non-inductive resistor mounted on a heat sink
C. A low-voltage power supply and a DC relay
D. A 50-ohm reactance used to terminate a transmission line

A dummy load.

A dummy load is a necessary accessory for any ham shack, whether you buy one or build one. Extended tuning up on the air is one of the infallible hallmarks of a "lid." Learn to use a dummy load; don't be a dummy load! ***A dummy load consists of a non-***

inductive resistor of sufficient power rating to handle your transmitter power and a heat sink to handle the heat of the non-radiated power. Classic dummy loads such as the Heath "Cantenna" are available at nearly every swap meet and can handle a "full gallon" (FCC 1,500 watts legal limit) of power for a while. **ANSWER B.**

Website Resources

IF YOU'RE LOOKING FOR	THEN VISIT
JPOLE Antennas	www.jpole-antenna.com
Portable Antennas	www.buddipole.com
VHF/UHF Beams/Verticals	www.arrowantennas.com
VHF/UHF Antennas	www.aorusa.com/antennas
	www.cometantenna.com
	www.Cushcraft.com
	www.diamondantenna.net
	www.directivesystems.com
	www.M2INC.com
	www.MFJEnterprises.com
	www.mosley-electronics.com
	www.NEW-Tronics.com
	www.pulseelectronics.com
	www.Radioworks.com
The ARRL Antenna Book	www.arrl.org/shop/Antennas
Required reading for every ham	www.goodreads.com/bookshow/851354.Practical_Antenna_Handbook_With_CDROM_
This is one of the best sites for learning about receiving antennas	http://www.hard-core-dx.com/nordicdx/antenna/
This is a portal to many other links of interest to DXing (long distance HF operating). Start out with the Antennas tab for band-specific antenna ideas.	https://www.dxzone.com/
Connectors	www.QSRadio.com

Watch This! **https://www.youtube.com/playlist?list=PL1KAjn5rGhizk5f2_whBLJaOHj59OCI_3 (video 272)**

Safety First!

T0A06 What is a good way to guard against electrical shock at your station?

A. Use three-wire cords and plugs for all AC powered equipment
B. Connect all AC powered station equipment to a common safety ground
C. Install mechanical interlocks in high-voltage circuits
D. All these choices are correct

All of these are good wiring techniques that will keep you safe around some of your larger ham radio equipment as you upgrade all the way to Amateur Extra Class. **ANSWER D.**

T0A03 In the United States, what circuit does black wire insulation indicate in a three-wire 120 V cable?

A. Neutral
B. Hot
C. Equipment ground
D. Black insulation is never used

The black wire insulation in a three-wire 120 volt AC cable is the HOT wire! It is usually the one with a switch and a fuse inside the equipment. Black is HOT! Do not confuse this with the red and black wires for a DC circuit in your car! **ANSWER B.**

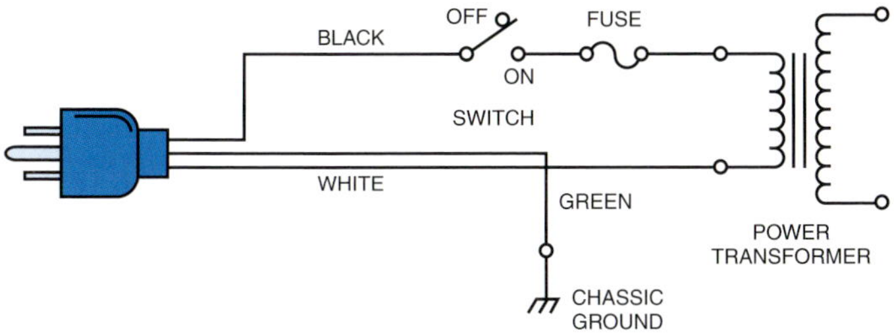

AC Line Connections

T0A08 Where should a fuse or circuit breaker be installed in a 120V AC power circuit?

A. In series with the hot conductor only
B. In series with the hot and neutral conductors
C. In parallel with the hot conductor only
D. In parallel with the hot and neutral conductors

In a home circuit, ***the circuit breaker or fuse should only be in the hot (line) side of the circuit***. The neutral should always be connected for the sake of safety, and it should always be at the same potential as the *safety ground.* Under normal operating conditions, no current should ever flow through the safety ground. **ANSWER A**

T0A04 What is the purpose of a fuse in an electrical circuit?

A. To prevent power supply ripple from damaging a component
B. To remove power in case of overload
C. To limit current to prevent shocks
D. All these choices are correct

Fuses provide an "on-purpose" weak link to protect the system from too much current passing through. The weak ***fuse element will melt*** at a specific current and interrupt the ***circuit.*** **ANSWER B.**

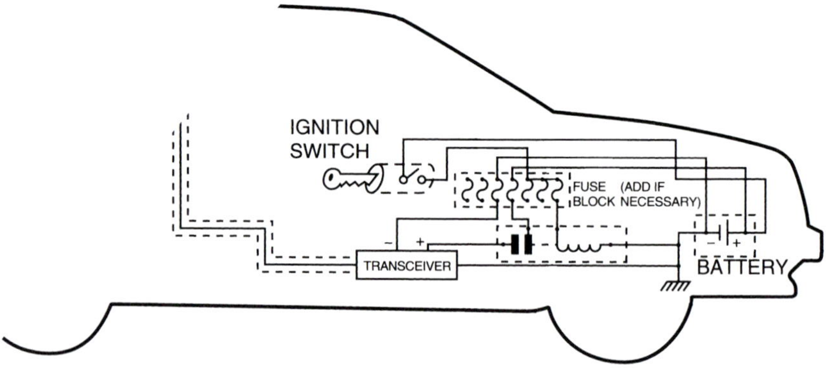

Place the fuses as close to the battery as possible

T0A05 Why should a 5-ampere fuse never be replaced with a 20-ampere fuse?

A. The larger fuse would be likely to blow because it is rated for higher current
B. The power supply ripple would greatly increase
C. Excessive current could cause a fire
D. All these choices are correct

A fuse is installed in both the red and black power leads of your transceivers to protect the wires from overload. Pulling too much current through the wires could cause them to heat up, and for the insulation to give off toxic smoke and eventually ***burst into flames***. If you ***substitute a 20-amp fuse for a 5-amp fuse*** on a small radio that is malfunctioning and blowing the 5-amp fuse, the 20-amp fuse might carry the load, causing the wires to heat up and, poof, you smoke your installation and worse, your entire shack. **ANSWER C.**

T0A11 What hazard exists in a power supply immediately after turning it off?

A. Circulating currents in the dc filter
B. Leakage flux in the power transformer
C. Voltage transients from kickback diodes
D. Charge stored in filter capacitors

High-voltage ***capacitors*** will many times ***store a charge*** for quite a while after the equipment has been turned off. Keep your fingers well away from any power supply until you have metered it for lethal voltages that may still be present, even though it has been unplugged from the wall for many hours. Always have a "chicken stick" available for discharging any capacitors in case the bleeder resistors fail. Always assume there are no safety bleeder resistors; better safe than toast! **ANSWER D.**

T0A12 Which of the following precautions should be taken when measuring high voltages with a voltmeter?

A. Ensure that the voltmeter has very low impedance
B. Ensure that the voltmeter and leads are rated for use at the voltages to be measured
C. Ensure that the circuit is grounded through the voltmeter
D. Ensure that the voltmeter is set to the correct frequency

High voltage is not something you need to cower in terror over, but you do need to respect it. Never work on high voltage alone. Use the "one hand method" to manipulate the probes of your voltmeter. Put your left hand in your back pocket. There is a large artery, full of highly conductive saltwater – newly oxygenated blood – that runs from your left hand to your heart, which you do not want to be in the equation. You are far less likely to blow out an aorta with a shock to your right hand. Better yet, it's best not to get any electrical shock at all. To be really safe, unplug the equipment in question, fully discharge any capacitors with a shorting stick, attach your ***voltmeter probes*** to the test points in question (with insulated alligator clips, if necessary) and then re-apply the power from a safe distance. This way you won't even be touching the voltmeter probes. This is how broadcast engineers measure 25 kilovolt DC plate circuits without dying. **ANSWER B.**

When working with 110 VAC circuits, be aware that common house voltage can kill! Be safe around any voltage level.

T0A02 What health hazard is presented by electrical current flowing through the body?

A. It may cause injury by heating tissue
B. It may disrupt the electrical functions of cells
C. It may cause involuntary muscle contractions
D. All these choices are correct

Never work around anything electrical in your bare feet. Put your shoes on! Never do your laundry with bare feet. If current is allowed to enter your finger and pass through your body out to ground, you will be toasted for sure. Never, ever work on anything electrical in bare feet! Look at ***answers A, B, and C***. Disgusting, aren't they? Indeed, current flowing through your body is a lethal health hazard! **ANSWER D.**

T0A10 What hazard is caused by charging or discharging a battery too quickly?

A. Overheating or out-gassing
B. Excess output ripple
C. Half-wave rectification
D. Inverse memory effect

Overcharging any type of battery, including your little handheld battery, could cause it to ***heat up*** and ***give off dangerous gas***, or even ***explode***. Be very cautious even with your new handheld battery and only charge it with the supplied charger or with a charger from a recommended ham radio battery charger company. **ANSWER A.**

T0A01 Which of the following is a safety hazard of a 12-volt storage battery?

A. Touching both terminals with the hands can cause electrical shock
B. Shorting the terminals can cause burns, fire, or an explosion
C. RF emissions from a nearby transmitter can cause the electrolyte to emit poison gas
D. All these choices are correct

Storage batteries can supply immense amounts of current and can act like an arc welder if their terminals are shorted out. Although there are several other dangers associated with storage batteries, ***high current shorts*** rank at the top of things to avoid. Be careful when replacing automotive batteries, as well – dropping a wrench across the battery terminals ***can cause immediate and extensive hissing and fizzing***. **ANSWER B.**

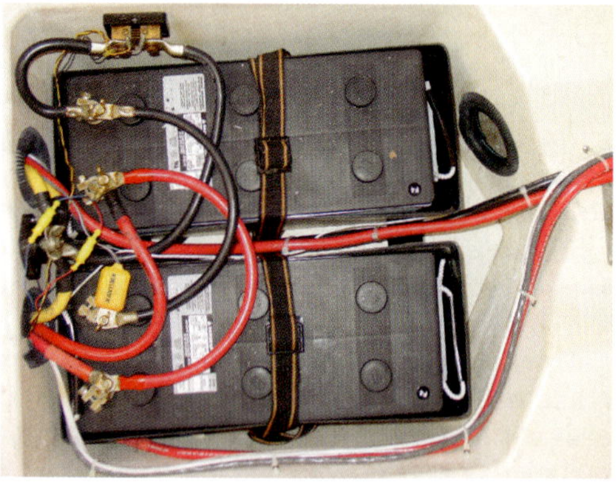

Recreational vehicle storage batteries, in parallel

T0B04 Which of the following is an important safety precaution to observe when putting up an antenna tower?

A. Wear a ground strap connected to your wrist at all times
B. Insulate the base of the tower to avoid lightning strikes
C. Look for and stay clear of any overhead electrical wires
D. All these choices are correct

Anytime you're working on an antenna system, look all over the place for overhead electrical wires and steer clear of these fatal attraction conductors. Normally those wires on the top of a power pole are uninsulated bare wire, and a brush with those 2 top wires could be your last day in ham radio. ***Watch out for overhead electrical wires!*** To see for yourself what can go wrong, search for Power Line Safety Video on YouTube.com. Some of these are startling. **ANSWER C.**

T0B06 What is the minimum safe distance from a power line to allow when installing an antenna?

A. Add the height of the antenna to the height of the power line and multiply by a factor of 1.5
B. The height of the power line above ground
C. 1/2 wavelength at the operating frequency
D. Enough so that if the antenna falls, no part of it can come closer than 10 feet to the power wires

Make sure your antenna system cannot accidentally fall close to or on power lines. A downed antenna system should be at least ***10 feet away from any nearby wires***. Your first antenna is likely to be a dual-band, white fiberglass 10' lightweight mast antenna. But you don't need to mast-mount this antenna. As long as you can get it up on the roof, you're good to go with exquisite range. Normally the mast antenna is affixed to a vent pipe. Run good coax, the size of your thumb. No sissy stuff the size of your little finger. Attach this antenna to your two-band portable or mobile rig downstairs and enjoy some great conversations. **ANSWER D.**

Stay well away from power lines!

T0B09 Why should you avoid attaching an antenna to a utility pole?

A. The antenna will not work properly because of induced voltages
B. The 60 Hz radiations from the feed line may increase the SWR
C. The antenna could contact high-voltage power lines
D. All these choices are correct

Never ever mess with any wire attached to a phone pole, utility pole, or power pole. Stay away from any type of utility pole as likely the high voltage wires on the very top could have some leakage that could ruin not only your equipment but you for good. ***Watch out for high voltage power wires!*** **ANSWER C.**

T0B02 What is required when climbing an antenna tower?

A. Have sufficient training on safe tower climbing techniques
B. Use appropriate tie-off to the tower at all times
C. Always wear an approved climbing harness
D. All these choices are correct

Double check that your ***fall prevention safety harness*** is OSHA-approved and that all connections are cinched up before climbing the tower. Always have a fellow ham help you from ground level. Never climb a tower alone! Wear ***safety goggles and a hard-hat***, and always observe climbing safety every step of the way up and down. Always double up on your safety straps and tie offs once you begin working on your antenna system up on the tower. **ANSWER D.**

Always wear a hard hat and safety glasses when working on an antenna tower.

T0B03 Under what circumstances is it safe to climb a tower without a helper or observer?

A. When no electrical work is being performed
B. When no mechanical work is being performed
C. When the work being done is not more than 20 feet above the ground
D. Never

Any time you are working on the roof or on an antenna tower, make sure you have your helpers and observers adorned with safety glasses and hard hats. ***Never*** work on anything electrical or mechanical, or more than 2 inches off the ground, without a helper nearby. **ANSWER D.**

T0B05 What is the purpose of a safety wire through a turnbuckle used to tension guy lines?

A. Secure the guy line if the turnbuckle breaks
B. Prevent loosening of the turnbuckle from vibration
C. Provide a ground path for lightning strikes
D. Provide an ability to measure for proper tensioning

It's an inviolable law of nature than any mechanical fastener will loosen with time. Turnbuckles will "un-turn" if there's no deliberate means applied to prevent this. ***Safety wire should be inserted through turnbuckles to prevent loosening***. Nuts and bolts subject to movement such as tilt-over tower hinges should be fitted with safety wire as well. **ANSWER B.**

T0B07 Which of the following is an important safety rule to remember when using a crank-up tower?

A. This type of tower must never be painted
B. This type of tower must never be grounded
C. This type of tower must not be climbed unless it is retracted, or mechanical safety locking devices have been installed
D. All these choices are correct

Never, ever climb a crank-up tower if it is partially or fully cranked up. Crank it down! Crank up towers rely on the stainless-steel cable to keep them at their "ultimate high." If you climb a crank up fully extended, you could overload the cable, and if the cable snaps, both you and the upper sections of the tower will nosedive to the ground. Never ever climb a crank up when it is cranked up! Crank-up towers are guillotines in disguise; if anything fails, you could easily end up beside yourself! **ANSWER C.**

If you climbed a crank-up tower and it broke and collapsed, you could lose an arm or a leg. Crank it down!

T0B11 Which of the following establishes grounding requirements for an amateur radio tower or antenna?

A. FCC Part 97 rules
B. Local electrical codes
C. FAA tower lighting regulations
D. UL recommended practices

Before erecting an antenna always check the ***local electrical codes***, which address tower grounding and antenna installation. Most local electric codes closely follow the National Electric Code (NEC) with regard to tower grounding and other related antenna issues but may impose additional requirements above and beyond the NEC in some localities. You must also, of course, comply with Part 97 rules and FAA tower height and lighting regulations where applicable, but these do not address grounding requirements. An excellent and comprehensive resource on antenna grounding and related issues is the ARRL book Grounding and Bonding for the Radio Amateur. **ANSWER B.**

T0B08 Which is a proper grounding method for a tower?

A. A single four-foot ground rod, driven into the ground no more than 12 inches from the base
B. A ferrite-core RF choke connected between the tower and ground
C. A connection between the tower base and a cold water pipe
D. Separate eight-foot ground rods for each tower leg, bonded to the tower and each other

Each tower leg – usually three – ***gets its own 8' ground rod***. Each rod is bonded to each other forming a triangle, then bonded to the tower for lightning protection. **ANSWER D.**

T0A09 What should be done to all external ground rods or earth connections?

A. Waterproof them with silicone caulk or electrical tape
B. Keep them as far apart as possible
C. Bond them together with heavy wire or conductive strap
D. Tune them for resonance on the lowest frequency of operation

Bonding all external ground connections together provides a number of benefits. The primary benefit is that the electrical potential of every piece of equipment is the same, eliminating the possibility of electrical shock by coming in contact with different units. It also provides some protection from induced field damaged caused by nearby lightning strikes. There isn't much you can do in the case of a direct lightning strike except to duck! Effective lightning protection attempts to avoid direct hits, rather than dealing with them after the fact. A solid grounding scheme is a large part of an effective lightning protection plan. Incidentally, you may have heard it said that lightning "follows the path of least resistance." This is wrong; in reality, lightning follows all possible paths! **ANSWER C.**

T0B01 Which of the following is good practice when installing ground wires on a tower for lightning protection?

A. Put a drip loop in the ground connection to prevent water damage to the ground system
B. Make sure all ground wire bends are right angles
C. Ensure that connections are short and direct
D. All these choices are correct

Keep all tower ground leads as short as possible to the tower earth grounding system. Avoid sharp turns, as a lightning strike needs a short straight direct path to the soil ground system of multiple ground rods pounded in the earth, right at the tower foundation. **ANSWER C.**

T4A08 Which of the following conductors is preferred for bonding at RF?

A. Copper braid removed from coaxial cable
B. Steel wire
C. Twisted-pair cable
D. Flat copper strap

The best method of grounding your equipment is with flat copper foil ribbon. This is available from most ham radio stores.

The ***flat strap*** offers the best surface area to bleed off static and to minimize ground currents that could cause interference. The strap usually comes in 3 inch widths, and you can fold it once or twice in order to snake it down to a healthy ground rod. **ANSWER D.**

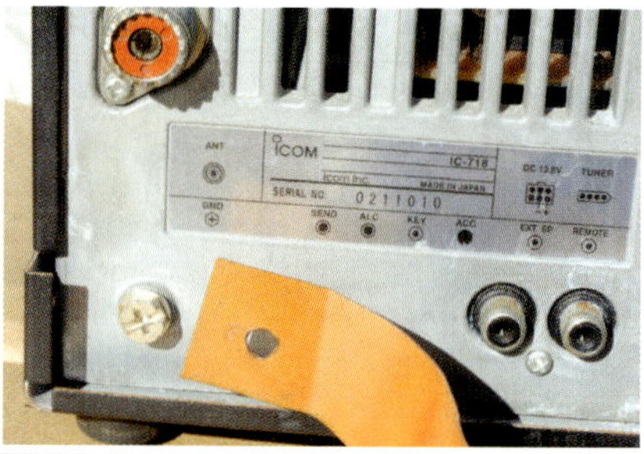

Copper Foil Ground Strap Provides Good Surface Area Ground

T0B10 Which of the following is true when installing grounding conductors used for lightning protection?

A. Use only non-insulated wire
B. Wires must be carefully routed with precise right-angle bends
C. Sharp bends must be avoided
D. Common grounds must be avoided

When you run your grounding conductors, ***avoid sharp bends*** as you don't want to confuse a lightning bolt on how to get from here to there directly. No bends. **ANSWER C.**

T0A07 Where should a lightning arrester be installed in a coaxial feed line?

A. At the output connector of a transceiver
B. At the antenna feed point
C. At the ac power service panel
D. On a grounded panel near where feed lines enter the building

A lightning arrestor is usually a spark gap of some sort (sometimes using actual spark plugs) to direct lightning induced electric arcs from a transmission line to ground. These are available in coaxial and balanced line configurations. The most effective location for a coaxial lightning arrestor is ***just before the spot where the coaxial cable enters the building, through a properly grounded panel***. Coaxial lightning arrestors usually have a gas filled arc gap. **ANSWER D.**

T0C04 What factors affect the RF exposure of people near an amateur station antenna?

A. Frequency and power level of the RF field
B. Distance from the antenna to a person
C. Radiation pattern of the antenna
D. All these choices are correct

The term "RF" refers to "radiofrequency" energy that comes off of your ham radio antenna system on transmit. The radiofrequency energy can be analyzed as a radiofrequency "field," and these "RF fields" must be evaluated to determine the environmental effects that this radio energy may have on our well-being. The important factors in evaluating the environmental effects of RF emissions are: frequencies and power levels of your transmitter; how high up your antenna is and how far it is away from people; the radiation pattern of your antenna; and, to some extent, ground reflections of the radiated energy. ***All of these factors*** are important to consider. **ANSWER D.**

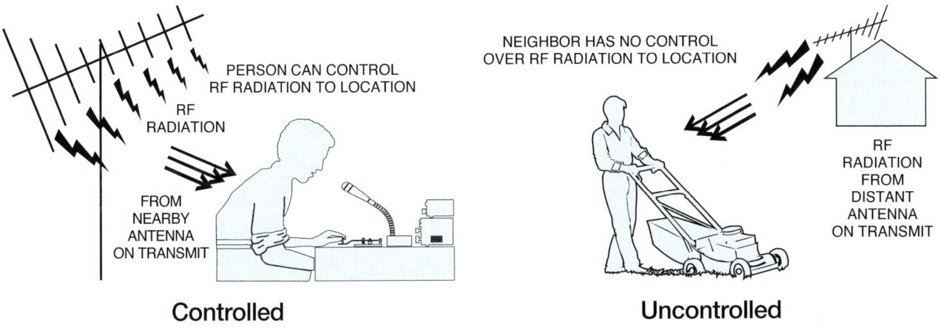

RF Radiation Exposure Environments

T0C05 Why do exposure limits vary with frequency?

A. Lower frequency RF fields have more energy than higher frequency fields
B. Lower frequency RF fields do not penetrate the human body
C. Higher frequency RF fields are transient in nature
D. The human body absorbs more RF energy at some frequencies than at others

Remember that your operation on different radio bands changes the frequency of your RF emissions. ***At certain wavelengths*** around 6 meters, ***the human body absorbs more RF energy***. On lower-frequency worldwide wavelengths, the body absorbs less RF energy. **ANSWER D.**

T0C02 At which of the following frequencies does maximum permissible exposure have the lowest value?

A. 3.5 MHz
B. 50 MHz
C. 440 MHz
D. 1296 MHz

We call "maximum permissible exposure" MPE. Every radio frequency has different effects on the body, but the common Technician Class ham radio 6-meter band, at ***50 MHz***, requires added distance between you and the transmitting antenna. As long as your 6-meter antenna is more than 20 feet away from you, at modest power levels, you're going to live to a ripe old age without problems from your radio hobby. Never put any type of ham radio antenna, other than the little antenna on your handheld, within 5 feet of your smiling face or body. **ANSWER B.**

T0C03 How does the allowable power density for RF safety change if duty cycle changes from 100 percent to 50 percent?

A. It increases by a factor of 3
B. It decreases by 50 percent
C. It increases by a factor of 2
D. There is no adjustment allowed for lower duty cycle

This is about as straightforward a calculation as you can get. Just calculate the duty cycle (ratio of ON time to OFF time) and multiply that by the exposure during the ON time to get your total exposure time. 6 minutes divided by 3 minutes = 2, ***twice the power*** density. **ANSWER C.**

T0C06 Which of the following is an acceptable method to determine whether your station complies with FCC RF exposure regulations?

A. By calculation based on FCC OET Bulletin 65
B. By calculation based on computer modeling
C. By measurement of field strength using calibrated equipment
D. All these choices are correct

You can determine how your station complies with FCC RF exposure regulation by using The W5YI RF Safety Tables in the Appendix of this book. You can use the tables to estimated safe distances based on FCC OET Bulletin No. 65, or by your own calculations based on computer modeling. You also can actually go out there with a field-strength meter and measure the power density levels making sure to use calibrated equipment. ***All of these choices*** are a good way to determine whether or not you are going to expose yourself or your neighbors unnecessarily. **ANSWER D.**

T0C08 Which of the following actions can reduce exposure to RF radiation?

A. Relocate antennas
B. Relocate the transmitter
C. Increase the duty cycle
D. All these choices are correct

In order to prevent exposure to RF radiation in excess of the FCC-specified limits, you may need to ***relocate your antennas***. **ANSWER A.**

T0C01 What type of radiation are radio signals?

A. Gamma radiation
B. Ionizing radiation
C. Alpha radiation
D. Non-ionizing radiation

The transmission of radiofrequency energy from your antenna is considered ***"non-ionizing" radiation***. This is altogether different from X-ray, gamma ray, and ultraviolet radiation. **ANSWER D.**

T0C12 How does RF radiation differ from ionizing radiation (radioactivity)?

A. RF radiation does not have sufficient energy to cause chemical changes in cells and damage DNA
B. RF radiation can only be detected with an RF dosimeter
C. RF radiation is limited in range to a few feet
D. RF radiation is perfectly safe

The simplest definition of ionizing radiation is radiation that ionizes things – that is, causes chemical reactions, like that used for cancer radiation treatments. Unless you're transmitting ultraviolet, you probably aren't going to be significantly ionizing anything at ***amateur radio power levels***. As an amateur operating under legal power limits, your chances of causing any ***ionizing radiation are essentially zero***. **ANSWER A.**

T0C09 How can you make sure your station stays in compliance with RF safety regulations?

A. By informing the FCC of any changes made in your station
B. By re-evaluating the station whenever an item in the transmitter or antenna system is changed
C. By making sure your antennas have low SWR
D. All these choices are correct

As you develop your den into a full-fledged radio room, always evaluate and ***re-evaluate all of the equipment*** going in and all of the antennas above you, on the roof. Make sure you are not going to be transmitting and subjecting your family and neighbors to too much RF radiation. Follow the guidelines we have for you in this book and stay safe around radio waves. **ANSWER B.**

T0C11 What is the definition of duty cycle during the averaging time for RF exposure?

A. The difference between the lowest power output and the highest power output of a transmitter
B. The difference between the PEP and average power output of a transmitter
C. The percentage of time that a transmitter is transmitting
D. The percentage of time that a transmitter is not transmitting

It's generally assumed that, as a radio amateur, you're going to be doing a lot more listening than talking. Well, other than being polite and civilized, there's another good reason to listen more than you yack! The less you talk, the lower ***the percentage of time you are actually transmitting***. Your total RF exposure will be less, at least for a given amount of power. RF exposure at HF frequencies is generally of little significance, but at certain VHF frequencies, it can be something to at least think about. All things in moderation. **ANSWER C.**

T0C10 Why is duty cycle one of the factors used to determine safe RF radiation exposure levels?

A. It affects the average exposure to radiation
B. It affects the peak exposure to radiation
C. It takes into account the antenna feed line loss
D. It takes into account the thermal effects of the final amplifier

The amount of RF energy your body can safely tolerate is a function of both power density and time. If the power density is high, you can only tolerate short exposures. If the power density is low, you can safely tolerate much longer exposures. Remember that ENERGY is the product of Power times Time. It is the actual absorbed ENERGY that determines the safe level of exposure. The ratio of ON time to OFF time is the ***duty cycle***. The effective ENERGY exposure is the product of the power density and the duty cycle. If a transmitter is on for one minute and off for one minute, the duty cycle is 50%. This means that with a 50% duty cycle, you can tolerate twice the power density as you would if the transmitter were continuously on. Both conditions result in the same ENERGY. **ANSWER A.**

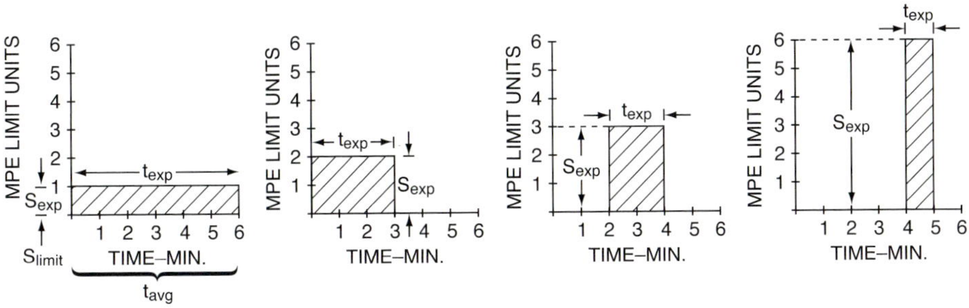

T0C13 Who is responsible for ensuring that no person is exposed to RF energy above the FCC exposure limits?

A. The FCC
B. The station licensee
C. Anyone who is near an antenna
D. The local zoning board

It is up to the station licensee to assure that no person is exposed to RF energy above the FCC exposure limits. Fortunately, at legal amateur power levels, it is difficult to exceed these limits, and at HF frequencies and lower, it is nearly impossible. The ARRL has just released a revised, detailed guidebook on how to determine these exposure limits for your station, and how to remedy them if you exceed them, or come uncomfortably close. **ANSWER B**

T0C07 What hazard is created by touching an antenna during a transmission?

A. Electrocution
B. RF burn to skin
C. Radiation poisoning
D. All these choices are correct

Your Technician Class license allows you to operate up to 1500 watts output of RF energy above 30 MHz. Make sure that no one can touch the transmitting antenna with this major amount of power on it because ***accidentally touching the antenna will cause someone to be burned***. **ANSWER B.**

Website Resources

▼ IF YOU'RE LOOKING FOR	▼ THEN VISIT
RF safety information	www.arrl.org/safety
	www.arrl.org/files/file/Technology/RFsafetyCommittee/7806011.pdf
Station grounding tips	www.w8ji.com/station_ground.htm

Watch This! **https://www.youtube.com/playlist?list=PL1KAjn5rGhizk5f2_whBLJaOHj59OCl_3 (videos 263, 344, + 265)**

5

Taking the Exam & Receiving Your First License

This chapter tells you how to find an exam session, how the examination will be given, who is qualified to give the Element 2 exam, and what happens after you complete it. There also are some good tips on finding a Volunteer Examiner (VE) team near you.

EXAMINATION ADMINISTRATION

All amateur radio service examinations are administered by licensed amateur operators who volunteer and who are accredited by a Volunteer Examiner Coordinator (VEC). Each exam session is coordinated by a national or regional VEC. Licensed amateurs who hold a General Class or higher class license may be accredited by a VEC to administer the Element 2 Technician Class examination. Advanced Class VEs may administer the Element 2 exam, and the Element 3 General Class exam. Extra Class VEs may administer all examinations, including the Element 4 Amateur Extra Class exam. A team of 3 or more officially-accredited VEs is required to form an examination session.

The Volunteer Examiners who will administer your exam session are hams, and as volunteers receive no pay for their time and skills. However, they are permitted to charge you a nominal fee for certain reimbursable expenses incurred in preparing, administering, and processing the examinations. The current fee is about $14.00.

HOW TO FIND AN IN-PERSON OR REMOTE ON-LINE EXAM SITE

In-person exam sessions are held regularly in communities all around the U.S. providing easy access at sites throughout the nation. The exam site could be a public library, a fire house, someone's office, or even in a private home. Normally, each examination team informs their sponsoring VEC with a schedule of their planned sessions, so the easiest way for you to find an exam session near you is to call a VEC or check their website. A complete list of VECs is located on page 218 in the Appendix.

Many VE teams now offer remote, on-line exam sessions via the internet. Go to the VEC website (see page 218), select the date and time for your remote exam session, then register for that session and pay the exam fee. Once you do so, you will receive an email confirming your on-line exam date and time, and a Zoom address for the remote on-line exam session.

Want to find a test site fast?
Visit the W5YI-VEC website at **www.w5yi.org**, or call 800-669-9594.

When you go on-line to the remote exam session, the first thing you will do is complete Form 605. Then the 35-question exam will appear on your screen. The exam is administered using ExamTools software that allows you to read the questions, see the answer choices, and select the correct answer. When you finish, the computer scores your exam and you know right away whether you passed.

When the session is over your exam results are sent electronically by the VE Team directly to the VEC office for proofing and processing to the FCC. License grants are normally given the same or next business day. Your entire on-line session is administered and observed by a team of 3 or more VEs from remote locations, just as if it were an in-person session.

Any of the VECs listed can provide you with the contact information of a local Volunteer Examiner team leader who can tell you the schedule of upcoming, in-person exam sessions near you. This may be given in a searchable list on their websites. To find a remote, on-line exam session, visit the VEC's website where you will find a link that shows the schedule of upcoming exam sessions.

Their websites will also tell give you what you will need to bring with you to the exam session, the amount of the current exam fee, and other important information.

ExamTools **was created in 2010 by Richard Bateman, KD7BBC, for use with a Volunteer Examiner team in Utah. The original goal was to reduce paperwork and provide computerized testing in a computer lab environment. Since then, it has been expanded significantly and is now backed by a thriving online community. It is the most-used system for administering remote, on-line exams with video conferencing software.**

Accommodating Persons with Disabilities

Volunteer Examiners are required to assist disabled examinees. The VEs must take into account the particular disability that the individual has and determine the proper way to accommodate the disability.

For example, VEs are permitted to read the questions and answer choices to a blind applicant and complete their answer sheet. If a blind applicant is given a question based on an illustration such as a schematic diagram, the VEs are allowed to substitute another question. If an examinee has hand coordination difficulties he/she can be accommodated by dictating their answers to an administering VE who will fill in the answer sheet.

Also, the VEs may defer the accommodating examination for that individual to a time and place when the VEs will be prepared to administer the exam. *If you are a disabled examinee, make sure to call the VE to make a reservation for an examination and tell them in advance of your disability and need for special accommodation.*

BEFORE THE EXAM GET YOUR FRN!

One of the first things you'll do when you get to your in-person or remote on-line exam session is complete FCC Form 605, the application for your Amateur Operator/Primary Station License. In order to complete Form 605 you need an FRN

FCC Registration Number. As of July 1, 2021, the FCC no longer allows applicants to use their Social Security Number, so **getting an FRN is mandatory**.

So what's an FRN?

An FRN is a 10-digit identification number assigned to an individual or a business registering with the FCC. This unique FRN is used to identify you and is used in lieu of your SSN to help prevent identity theft.

You should obtain your FRN prior to taking you exam. Visit the following website and follow the instructions to obtain your FRN:
apps.fcc.gov/cores/userlogin.do

Navigating the FCC Websites

There are two FCC websites that you will be visiting as a licensed Amateur Radio operator. The first is the CORES website, apps.fcc.gov/cores/userlogin.do. On this website home page you will find the following information to help you navigate this site. It includes a video tutorial to teach you how to use the many features of the site.

```
Welcome to the COmmission REgistration System (CORES).
If you DO NOT have an FCC Username, please click the "REGISTER"
button below to register an FCC Username.
CORES allows you to:
  • Register and Manage the 10 digit FCC Registration Number
    (FRN).
  • View financial standing (RED or Green Light Status) with the
    FCC and make payments.
  • Retrieve payments awaiting completion.
  • Enter and pay Application fees manually.
  • Retrieve and pay ULS fees.
  • File and pay annual Regulatory Fees.
  • Manage Incentive Auction financials.
Click Here for CORES Tutorial Videos.
News releases related to the FCC Registration Number.
```

The second website is for the Universal Licensing System (ULS). The FCC no longer mails paper licenses and only copies are available for download. Be sure to watch your inbox and spam box after your exam for an FCC email with the link to a printable copy of your new license. You can print a reference copy of your license by searching the FCC ULS database using your FRN.
Wireless2.fcc.gov/UlsApp/UlsSearch/searchLicense.jsp

WHAT TO BRING TO THE EXAM

Here's what you'll need to bring with you for your Technician Class Element 2 written examination:

1. Examination fee of approximately $14.00 in cash.
2. Your Federal Registration Number.
3. Personal photo identification card, such as your driver's license or student ID.
4. Some sharp pencils and fine-tip pens. Bring a backup!

5. Calculators may be used, so bring your calculator. (**NOTE:** Remote, on-line exam sessions do not allow the use of handheld calculators. You must use the calculator on your computer, tablet, or mobile phone.)
6. Any other item that the VE team asks you to bring.

COMPLETING NCVEC FORM 605

When you arrive at the exam site, one of the first things you will do is compete the NCVEC Form 605. This is your application for an FCC Amateur Radio license. When you pass your exam, your VE team forwards your Form 605 to their sponsoring Volunteer Examination Coordinator. The VEC verifies your exam result, transfers your information to an electronic file, and then submits it to the FCC's Licensing Bureau, which issues your Technician Class (or upgrade) license.

Your application may be delayed or kicked-back to you if the VEC can't read your writing. So make absolutely sure you print as legibly as you can, and carefully follow the instructions on the form.

Name

The FCC requires your legal birth name – your last name, first name, middle initial, and suffix such as junior or senior, etc. *You must stay absolutely consistent with your name on any future Form 605s for upgrades or changes of address.* Once the FCC issues your license, all information on the license remains as first submitted to the FCC until such time you change your mailing address, modify your license at time of renewal, or legally change your name due to marriage or other reasons.

NCVEC QUICK-FORM 605 APPLICATION
AMATEUR OPERATOR/PRIMARY STATION LICENSE

SECTION 1 - TO BE COMPLETED BY APPLICANT — PLEASE PRINT LEGIBLY!

PRINT LAST NAME	SUFFIX (Jr., Sr.)	FIRST NAME	M.I.	STATION CALL SIGN (IF ANY)
MAILING ADDRESS (Number and Street or P.O. Box)				FEDERAL REGISTRATION NUMBER (FRN)
CITY		STATE CODE	ZIP CODE (5 or 9 Numbers)	
DAYTIME TELEPHONE NUMBER (Include Area Code)		E-MAIL ADDRESS (MANDATORY TO RECEIVE LICENSE NOTIFICATION EMAIL FROM FCC)		

Basic Qualification Question: **ANSWER REQUIRED IN ORDER TO PROCESS YOUR APPLICATION**

Has the Applicant or any party to this application, or any party directly or indirectly controlling the Applicant, ever been convicted of a felony by any state or federal court? ☐ YES ☐ NO

If "YES", see "FCC BASIC QUALIFICATION QUESTION INSTRUCTIONS AND PROCEDURES" on the back of this form.

NCVEC Form 605

E-mail, Phone, and Mailing Address

Your e-mail address is very important! This is how the FCC will notify you of your new license and call sign under the Universal Licensing System. Once your license grant is awarded, the FCC will send you an e-mail with a link to your new or upgrade license. Since you probably haven't had e-mail exchanges with the FCC, make sure to watch you spam or junk folder for the important e-mail message.

The FCC has almost entirely stopped using snail-mail for its messages to amateur radio applicants and licensees. So once you have your FRN, if you ever change your e-mail address make sure you change it on the FCC registration site.

Provide your daytime telephone number in case the VEC or FCC needs to contact you. It probably is best to give them your cell phone number where they can most easily get in touch with you.

Fill in your mailing address – one that you plan to keep as permanent as possible. The FCC may use this address to contact you by mail, if necessary, so you need to keep it current on the FCC database along with you updated e-mail address.

Basic Qualifying Question

As of September, 2017, the FCC requires license applicants to answer a **Basic Qualifying Question**. It asks if you have every been convicted of a felony by any state or federal court. If you answer YES, then you are required to submit a statement directly to the FCC explaining the circumstances as well as copies of your court documents. Your written statement to the FCC should contain the reasons why you believe that it would be in the public interest to award you an amateur radio license notwithstanding the actual or alleged misconduct. DO NOT SEND THIS INFORMATION TO YOUR VEC.

Your statement and documentation must be submitted directly to the FCC within 14 days of your application. The VEC that submits your application to the FCC will also send you an email with your Application Number, which you must include with your statement. You can mail your statement and documentation to the FCC at:

FCC
1270 Fairfield Road
Gettysburg, PA 17325-7245

Or you can submit your statement and documentation electronically by sending it to: **attach605@fcc.gov**

Important Information about sending documents to the FCC: If you wish to insure that your documents sent to the FCC are kept private between you and the FCC, you must mark, stamp or write on each page of your documentation the word "CONFIDENTIAL." If you fail to do so, the FCC will post your documents to your license record for public viewing in the FCC ULS database.

Final Check

Once you're finished filling out your portion of the NCVEC Form 605, take a moment to double-check that your handwriting / printing is legible. If a single letter in your name can't be read clearly and is misinterpreted, subsequent electronic filings may be returned as "no action." Make sure your Form 605 is clear as a bell to your Volunteer Examiner team, who will then forward it to their VEC.

Your Examiners' Portion

Section 2 of the Form 605 is completed by your Volunteer Examiner team. The VEs will review your portion, making sure it is complete and legible, and when you pass your exam they will complete and sign their portion of the Form before sending it on to their Volunteer Examiner Coordinator.

Electronic Filing

After you pass your Element 2 Technician Class exam, your VE team will submit your amateur license application results to the VEC, either electronically or by snail mail. The VEC will certify your results and then electronically submit your application to the FCC. Within a day or two of having your application sent to the FCC, your call sign will be granted and it will show up on the FCC's ULS database. This is when you will receive that all-important e-mail announcing your new Technician Class, or upgrade, license!

TAKING THE EXAMINATION

There won't be any surprises on your upcoming Element 2 written exam. Every one of the 35 multiple choice questions on your exam will be taken from the 411 questions in this book. The wording of the questions and answer choices will be exactly as they appear here. The only possible change that the VEs are allowed to make is in the A B C D order of the answer choices.

So, get a good night's sleep before the exam day. If you're having a hard time with certain questions, make a list of them and work to memorize the answers, and review them just before you go into the test room. If someone is going with you to the session, have them ask you the questions so you can practice answering them. Listen to our audio theory CDs in your car as you drive to the session. Speed read keywords over and over again before the exam!

Check and Double-Check

When the VEs hand out the examination material, put your name, date, and exam number on the answer sheet. *Make no marks on the multiple-choice question sheet.* Only write on the answer sheet.

Read each question carefully. Take your time looking for the correct answer. Some answer choices start out looking correct, but end up wrong. *Don't speed read the questions and answer choices!*

When you are finished with the examination, go back over every question and double-check your answers. Read what you have selected as the correct answer and see if it agrees with the question.

W5YI-VEC

National Volunteer Examiner Coordinator

This certifies that:

NAME STATION CALL SIGN

DATE OF ISSUE: ____

CITY / STATE (Session Site)

NUMBER AND STREET

CITY STATE ZIP

has SUCCESSFULLY PASSED the following elements:

☐ Element 2, Technician Class ☐ Element 3, General Class ☐ Element 4, Amateur Extra Class

and will be given credit for this examination element when the appropriate additional examination element is passed at a subsequent examination session within 365 days from the date of issue of this certificate. *(See Section 97.505(a)6)).*

has SUCCESSFULLY PASSED all elements for the following operator license class:

☐ Technician Class Operator ☐ General Class Operator ☐ Amateur Extra Class Operator

If you already hold an FCC-issued amateur radio license, this certificate validates temporary (interim) operation with the rights and privileges of your new operator class (see Section 97.301 of the Commission's Rules) until you receive the license of your new operator class, or for a period not to exceed 365 days from the date of issue of this certificate. (See Section 97.9(b)). When operating on an interim basis in the telegraphy / data / image /RTTY mode, you must append your call sign with /KT (Technician), /AG (General) or /AE (Amateur Extra Class). Use the word "Temporary" before the identifier (KT, AG or AE) when operating in the voice mode. *(See Section 97.119(f)).*

THIS CERTIFICATE IS NOT A LICENSE PERMIT OR ANY OTHER KIND OF OPERATING AUTHORITY

SAMPLE

Sample CSCE.

When you finish, turn in all of your paperwork. Tell the VEs how much you appreciate their efforts to help promote ham radio. If you are the last one in the room, volunteer to help the VEs take down the chairs and tables.

And now, wait patiently for the examiners to announce that you have passed your exam. Chances are they will greet you with a smile and your Certificate of Successful Completion of Examination. Make sure to immediately sign this certificate when it is handed to you.

CONGRATULATIONS! YOU PASSED!!

After you pass the examination, CONGRATULATIONS and a big welcome to Technician Class privileges are in order! The world of VHF/UHF and microwave operating awaits you. And welcome to the additional high frequency privileges and the world of long-rang, HF operation that you earned. You can begin operating as soon as you obtain your official FCC call sign.

Depending upon how your Volunteer Examiner team files your NCVEC Form 605 with their sponsoring VEC, you might receive your license grant from the FCC in as little as a few days. This is true if the VE team sends your Form 605 to the VEC electronically. On the other hand, if your VE team sends your Form 605 to their VEC via snail mail, it could take a couple of weeks for your new license to appear on the FCC's ULS database.

Many W5YI VEC examination teams submit their paperwork electronically to the VEC office, and if you test with one of these VE teams, you will get your new license very quickly.

FCC NOW CHARGES LICENSING FEE

In April 2022, the FCC began charging a new $35.00 application fee for Amateur Radio licenses. Previously, licenses were granted free of charge. This FCC fee is in addition to the examination fee paid to the VE team.

When the FCC receives your examination information from the VEC and approves the application, you will receive an email with a link to the FCC Fee Payment Portal and payment instructions. You will pay the fee using your credit or debit card.

You have 10 days from the date of the email to pay the fee. Make sure to monitor your e-mail inbox and spam folder so you can pay your application fee promptly.

Alternatively, if you are unable to monitor your email inbox or spam folder, you can go on the FCC website **wireless2.fcc.gov/UlsApp/UlsSearch/searchLicense.jsp** and search for your approved application using your FRN number. When you see your pending application that has been approved by the FCC there, you can follow the instructions and to go to the FCC Fee Payment Portal to pay the $35.00 fee.

Look for the box that says Pending Applications

PA = Pending Application(s)
TP = Termination Pending
L = Lease

Page 1

Name	FRN	Radio Service	Status	Expiration Date
CLARK, MALACHI D	0031921539	HA	Active	02/15/2032
Name	**FRN**	**Radio Service**	**Status**	**Expiration Date**

Your ULS FRN search on the FCC website will look like this.

Either way, once your fee is paid, the FCC will issue your license grant, usually within a few hours on normal business days.

Next, you will receive a second email from the FCC with a link directly to your official license record. ***The link is good for 30 days.*** You also will be able to view, print, and download an official copy of your license by logging into your FCC ULS account.

The FCC discontinued mailing or providing printed licenses in February 2015. Licensees can log into ULS with their FRN and password at any time to view and manage their license record, including updating their email or mailing address.

YOUR FIRST CALL SIGN

To find your new call sign on the internet, go to: **wireless.fcc.gov/uls**. This will take you to the FCC Wireless Telecommunications Bureau ULS home page. On the center of the screen you will find a Search area. Click on License Search then search by name entering your last name first, a comma, then your first name. You also can look for your new call sign on **www.qrz.com**

Paperless Licenses

As of February, 2015, the FCC no longer routinely issues/mails a paper license document to amateur radio licensees and other Wireless Telecommunications Bureau (WTB) licensed services. The FCC maintains that the official Amateur Radio license authorization is the electronic record that exists in its Universal Licensing System database. Licensees may access their current, official authorization, and print out an unofficial "reference copy" of their license using the FCC ULS License search feature at **wireless.fcc.gov/uls**.

However, when you receive your first e-mail notification of your new call sign from the FCC and log-in to your license record, you can print out an official copy of your license. This is a one-time opportunity to print a copy without it showing an over-print saying Reference Copy.

Amateur Call Signs

As an aid to the enforcement of the radio rules, transmitting stations around the world are required to identify themselves at regular intervals when they are on the air. That is the primary purpose of your call sign. It is a unique identifier that belongs only to you – no one else in the world has the same call sign as you. For many hams, their call sign is as important to them as their own name!

By international agreement, the prefix letters of a station's call sign identifies the country in which that station is authorized. The prefix letters allocated to the United States are AA through AL, KA though KZ, NA through NZ, and WA through WZ.

In the U.S., amateur stations are assigned call signs that start with AA, AL, K, N, or W followed by a single digit number that indicates their location in a specific geographic area. *Table 5-1* details these areas, and they are shown on the map in *Figure 5-1*.

The suffix letters assigned after the number indicate the specific amateur station.

So, call sign KB9SMG tells you that KB is in the U.S., 9 tells you the region within the U.S. where the station is located, and SMG assigns this to a single individual / station. On the DX airwaves, hams can readily identify the national origin of the ham signal they hear by its call sign prefix.

Your Amateur Operator / Primary Station call sign is issued by the FCC licensing facility in Gettysburg, PA, on a systematic basis after it receives your application information from the VEC. There are four call sign groupings – A, B, C, and D. Each grouping has a specific format, as follows:

- Group A call signs are issued to Extra Class licensees. They have a 1-by-2, 2-by-1, or a 2-by-2 format. W9MU is an example. Your first FCC sequentially-issued Extra Class call sign will by a 2-by-2 format.
- Group B call signs are issued to grandfathered Advanced Class licensees. They have a 2-by-2 format that begins with K, N, or W. WA6PT is an example.
- Group C call signs were issued to General and Technician Class licensees. However, because of the large number of Technician licenses issued since the inception of the "no-code" license in 1991, Group C call signs are no longer automatically assigned. So, in accordance with FCC rules, all new General and Technician Class operators are issued Group D call signs. You can apply for a 1-by-3 Group C vanity call sign.
- Group D call signs are issued to new Technician and General Class licensees. The have a 2-by-3 format. KB9SMG is an example.

Table 5-1. Call Sign Area for U.S. Geographical Areas

Call Sign Area No.	Geographical Area
1	Maine, New Hampshire, Vermont, Massachusetts, Rhode Island, Connecticut
2	New York, New Jersey, Guam and the U.S. Virgin Islands
3	Pennsylvania, Delaware, Maryland, District of Columbia
4	Virginia, North and South Carolina, Georgia, Florida, Alabama, Tennessee, Kentucky, Midway Island, Puerto Rico[1]
5	Mississippi, Louisiana, Arkansas, Oklahoma, Texas, New Mexico
6	California, Hawaii[2]
7	Oregon, Washington, Idaho, Montana, Wyoming, Arizona, Nevada, Utah, Alaska[3]
8	Michigan, Ohio, West Virginia, American Samoa
9	Wisconsin, Illinois, Indiana
0	Colorado, Nebraska, North and South Dakota, Kansas, Minnesota, Iowa, Missouri, Northern Mariana Island

[1] Puerto Rico also issued Area #3.
[2] Hawaii also issued Area #7.
[3] Alaska also issued Areas #1 through #0.

Once assigned, a call sign is never changed unless the licensee specifically requests a change – even if they move out of the U.S. FCC licensees residing in foreign countries must show a U.S. mailing address on their applications. When you upgrade your amateur license, you may change your call sign, or you may keep your current call sign. So, if you get a real neat call sign as your first one, you can elect to keep it all the way to the top Amateur Extra Class license.

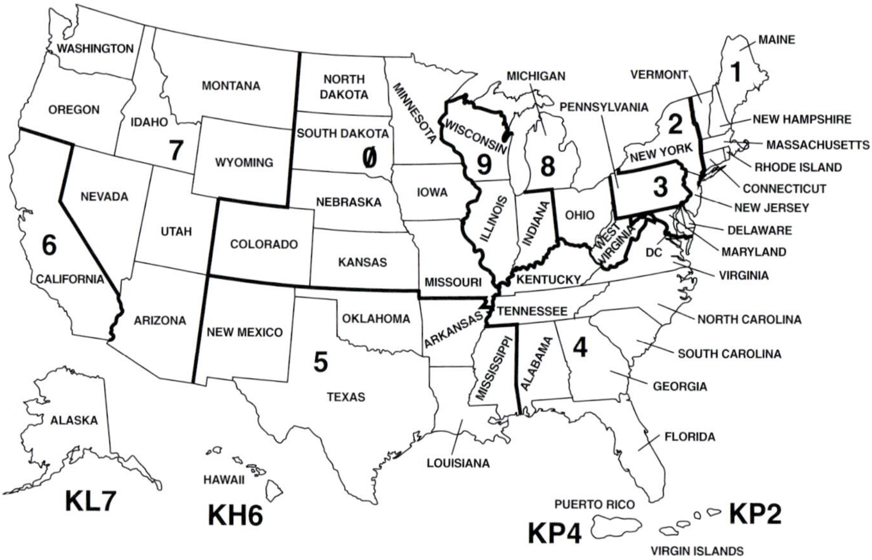

Figure 5-1. U.S. Call Sign Areas

So what does all this mean? You can no longer tell what grade of license someone has by the format of their call sign. And you can't tell what area of the country they are from by the number in their call sign – they may have moved from Maine to California. But you are only a few keystrokes away from being able to identify any ham by their call sign by visiting: **www.qrz.com**. This website maintains a database of call signs that you can search.

You can also look up call signs by the name of the licensee, or by the person's call sign by visiting the FCC's website at: **wireless.fcc.gov/uls**. Click on the search function and enter the call sign, or search by name.

Vanity Call Signs

Once you receive your first computer-assigned call sign, you can replace it with a vanity call sign of your choosing. This call sign could be made up of your initials, or represent your love of animals (K9DOG), or could be the call sign that your late mom or dad had when they were hams years ago.

Technician and General Class operators may request a vanity call sign from Group C or D. Group C call signs begin with the letter N, K, or W and have a 1-by-3 format, such as N8XYZ. Group D call signs have a 2-by-3 format, such as KA5GMO.

You can find out all the rules (and there are many!) regulating vanity call signs by visiting the W5YI Group website at: **www.w5yi.org** and clicking on the "Vanity Call Signs" button found on the menu on the left side of the screen.

You can find out how to apply for yourself by going to: **www.fcc.gov/common-filing-task-obtaining-vanity-call-sign.** While it is possible to apply for yourself, it isn't as easy as it sounds. But there is help available for a nominal fee.

Call Sign Services from W5YI

The W5YI VEC offers several call sign services for amateur radio operators.

To make it easier for you, and to help insure you get the vanity call sign you want, the W5YI Group will electronically file you application for you. For a nominal service fee, they will research if a specific call sign is available. You can enter your application and call sign choices right on their website at **www.w5yi.org**, or call them at 800-669-9594 if you have any questions.

In addition to their vanity call sign service, W5YI VEC also can:

- File for a change of address.
- File for name changes.
- Renew expiring licenses.
- Assign 1-by-1 special event call signs.
- Research former license records to confirm if you are eligible for paper upgrade to General Class without further examination, or verify your expired / former license privileges prior to your Technician examination to regain those privileges.
- File for new club station license, change of address, trustee or name of club on existing club station licenses and club vanity call signs.

Check their website at **www.w5yi.org** for more information on these and other services, or call 800-669-9594 during normal business hours.

THANK YOUR VE TEAM!

Your Volunteer Examiner team is made up of men and women who are volunteering their time and effort to provide you with examination opportunity. Your VE team spends many hours preparing for and administering the exam session, and completing and filing the session results with their VEC office. So please be sure to say "Thank You" to them when you complete your exam session.

And remember, while getting ready to take your exam, and all during the test session, follow the instructions of your VE team! If you're not clear about something, raise your hand and ask a question. If you're not sure about it, chances are others at the exam session have a similar question.

When you achieve the General Class or higher license level, it will be time for *you* to become a Volunteer Examiner! It will be one way in which you can help our hobby grow.

YOUR NEXT STEP

First, GET ON THE AIR as a new Technician Class operator. That's where you'll really begin your education as a ham radio operator. After you've spent some time operating on repeaters, and trying out skywaves on 6- and 10-meters, it will be time to consider upgrading to General Class, and then on to top license, Amateur Extra Class. Remember, no more code test, so all you'll need to upgrade are our *General Class* and *Extra Class books*.

SUMMARY

Welcome to the new, improved and simplified Amateur Radio service. There is no longer a Morse code test for any class of license. Restructured and simplified Element 2, Technician. Straight-forward Element 3, General Class. And for you high-techies, Element 4, the Amateur Extra Class.

Learn the code. Even though it's no longer requited, never in the history of ham radio has CW given you so many operating privileges on the worldwide bands as the new Technician Class license.

Become an *active* amateur, and help establish new radio clubs and volunteer examination programs near where you live. Introduce ham radio to kids, and let's keep our service growing!

FREE PASSING CERTIFICATE

When you pass your exams, I want to know about it! I have a very nice certificate available to you, suitable for framing plus free operating materials from ham equipment manufacturers. All I need is a large, self-addressed envelope with 12 first-class stamps on the inside to cover postage and handling and I'll send one your way. Write me at:

Gordon West Radio School
2414 College Drive
Costa Mesa, California 92626

Listen for me on the airwaves as WB6NOA. Say "hi" at many of the hamfests that I attend throughout the country every year. And if you ever would just like to speak with me, call me Monday through Thursday, 10 am to 4 pm (California time), 714-549-5000.

Welcome to ham radio! It's been **FUN** teaching you the Technician Class.

Gordon West
73's

Gordon West, WB6NOA

Congratulations! You passed!!

6

Learning Morse Code

On April 15, 2000, the FCC lowered the 20- and 13-word-per-minute Morse code requirements for worldwide frequency privileges for General and Extra Class operators down to 5 words per minute. On February 23, 2007, the FCC ***totally eliminated*** the Morse code test as a prerequisite for high frequency operation.

The elimination of the Morse code test for operation on worldwide frequencies conforms to international radio regulations. In 2003, the International Telecommunications Union (ITU) World Radiocommunication Conference voted to allow individual nations to determine whether or not to retain a Morse code test as a requirement to operate on frequencies below 30 MHz.

When the FCC eliminated the Morse code test for Technician Class operators in 1991 for VHF/UHF operating, the ruling was adopted with little opposition. However, the announcement that the FCC was considering total elimination of the Morse code test drew thousands of written comments to the FCC. Many comments supported code test elimination, while a minority urged the FCC to retain a code test because of the strong tradition of CW as a ham radio operating mode.

The Federal Communications Commission concluded that "...this change (eliminating the code test) eliminates an unnecessary burden that may discourage current amateur radio operators from advancing their skills and participating more fully in the benefits of amateur radio." The FCC Commissioners recognized the simple fact that learning Morse code was keeping many very technical, talented hams from obtaining their General and Extra Class licenses. Morse code is much like musical rhythms. Some people are tone deaf, and some people couldn't carry a rhythm in a hand basket.

So for years, the Morse code test was an insurmountable hurdle to many talented Technician Class hams who wanted to upgrade. If I could take these Techs and put them into one of my regular Morse code classes, we usually could get the majority of them through the CW test with outside home study, on the air practice (on 2 meters), and classroom study followed by the code test. But throughout the country, Morse code classes were few and far between and it is tough to learn new music and a new language without classroom instruction.

Technician Class operators could not practice on the worldwide airwaves to learn the code because these bands were reserved for only those operators who had already passed the code test. Running Morse code practice on a local 2 meter repeater was one option, but nothing beats the excitement of practicing code on the worldwide bands and hooking up with another station thousands of miles away.

As of February 23, 2007, we can now take new operators and introduce them to Morse code on the exciting worldwide bands!

Morse code is the ham radio operator's most basic language of short and long sounds, dits and dahs, or dots and dashes. Sailors have pounded SOS when trapped beneath a sailboat hull. In submarines, the tapping of Morse code gets the message through when there is no other way to communicate. Prisoners of war have tapped out Morse code messages, or BLINKED the code when being publicly displayed on television.

Ham operators use the code to get through when noise would otherwise cover up data or voice signals. Years ago, before road rage, fellow hams driving might greet each other by sending on their car horns H-I, a friendly salute to another ham.

So I encourage you to learn code. It is best mastered by sound along with memorizing the Morse code patterns seen on the upcoming pages. The pages show the number of dots and dashes to learn for a specific character, and learning the sound (rhythm) of Morse code is always the best way to practice. Let's see what these short sounds and long sounds are all about.

LOOKING AT MORSE CODE

The International Morse code, originally developed as the American Morse code by Samuel Morse, is truly international — all countries use it, and most commercial worldwide services employ operators who can recognize it. It is made up of short and long duration sounds. Long sounds, called "dahs," are three times longer than short sounds, called "dits." *Figure 6-1* shows the time intervals for Morse code sounds and spaces. *Figure 6-3,* on the next page, indicates the sounds for all the CW characters and symbols.

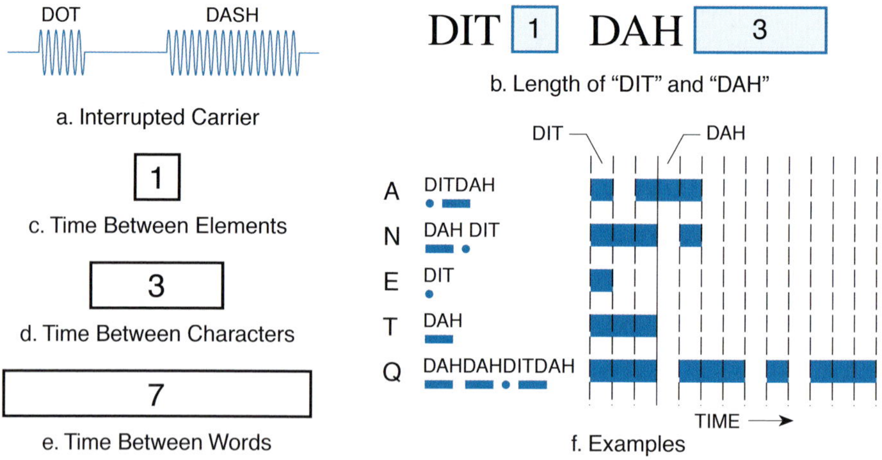

Figure 6-1. Time Intervals for Morse Code

a. Alphabet

LETTER	Composed of:	Sounds like:	LETTER	Composed of:	Sounds like:
A	• –	didah	N	– •	dahdit
B	– • • •	dahdididit	O	– – –	dahdahdah
C	– • – •	dahdidahdit	P	• – – •	didahdahdit
D	– • •	dahdidit	Q	– – • –	dahdahdidah
E	•	dit	R	• – •	didahdit
F	• • – •	dididahdit	S	• • •	dididit
G	– – •	dahdahdit	T	–	dah
H	• • • •	didididit	U	• • –	dididah
I	• •	didit	V	• • • –	didididah
J	• – – –	didahdahdah	W	• – –	ditdahdah
K	– • –	dahdidah	X	– • • –	dahdididah
L	• – • •	didahdidit	Y	– • – –	dahdidahdah
M	– –	dahdah	Z	– – • •	dahdahdidit

b. Special Signals and Punctuation

CHARACTER	Meaning:	Composed of:	Sounds like:
$\overline{AR}$	(end of message)	• – • – •	didahdidahdit
K	invitation to transmit (go ahead)	– • –	dahdidah
$\overline{SK}$	End of work	• • • – • –	didididahdidah
$\overline{SOS}$	International distress call	• • • – – – • • •	didididahdahdahdididit
V	Test letter (V)	• • • –	didididah
R	Received, OK	• – •	didahdit
$\overline{BT}$	Break or Pause	– • • • –	dahdidididah
$\overline{DN}$	Slant Bar	– • • – •	dahdididahdit
$\overline{KN}$	Back to You Only	– • – – •	dahdidahdahdit
Period		• – • – • –	didahdidahdidah
Comma		– – • • – –	dahdahdididahdah
Question mark		• • – – • •	dididahdahdidit
@	For Web Address	• – – • – •	didahdahdidahdi

c. Numerals

NUMBER	Composed of:	Sounds like:
1	• – – – –	didahdahdahdah
2	• • – – –	dididahdahdah
3	• • • – –	didididahdah
4	• • • • –	dididididah
5	• • • • •	dididididit
6	– • • • •	dahdidididit
7	– – • • •	dahdahdididit
8	– – – • •	dahdahdahdidit
9	– – – – •	dahdahdahdahdit
Ø	– – – – –	dahdahdahdahdah

Figure 6-3. Morse Code and Its Sound

CODE KEY

Morse code is usually sent by using a code key. A typical one is shown in *Figure 6-2a.* Normally it is mounted on a thin piece of wood or plexiglass. Make sure that what you mount it on is thin; if the key is raised too high, it will be uncomfortable to the wrist. The correct sending position for the hand is shown in *Figure 6-2b.*

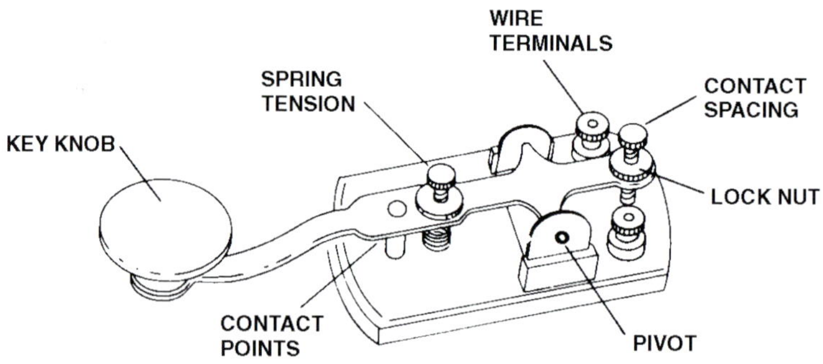

Figure 6-2a. Code Key

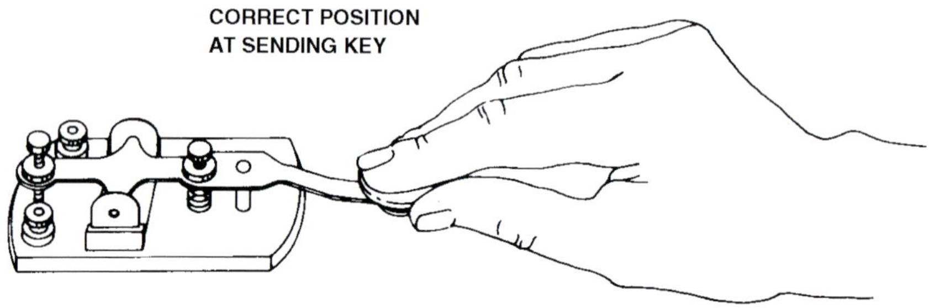

b. Sending Position

Figure 6-2b. Code Key for Sending Code

LEARNING MORSE CODE

The reason you are learning the Morse code is to be able to operate all modes on the worldwide bands—including CW. Here are five suggestions (four serious ones) on how to learn the code:

1. Memorize the code from the code charts in this book.
2. Use my fun audio course available at all ham radio stores, and from the W5YI Group.
3. Go out and spend $700 and buy a worldwide radio, and listen to the code live and on the air. You don't need to spend that much, but you can listen to Morse code practice on the air, as shown in *Table 6-1*.
4. Use a code key and oscillator to practice sending the code. Believe it or not, someday you're actually going to do code over the live airwaves, using this same code key hooked up to your new transceiver.
5. Play with code programs on your computer, and *have fun!*

Table 6-1. Radio Frequencies and Times for Code Reception

Pacific	Mountain	Central	Eastern	Mon.	Tue.	Wed.	Thu.	Fri.
6 a.m.	7 a.m.	8 a.m.	9 a.m.		Fast Code	Slow Code	Fast Code	Slow Code
7 a.m. – 1 p.m.	8 a.m. – 2 p.m.	9 a.m. – 3 p.m.	10 a.m. – 4 p.m.	**VISITING OPERATOR TIME**				
1 p.m.	2 p.m.	3 p.m.	4 p.m.	Fast Code	Slow Code	Fast Code	Slow Code	Fast Code
2 p.m.	3 p.m.	4 p.m.	5 p.m.	Code Bulletin				
3 p.m.	4 p.m.	5 p.m.	6 p.m.	Digital Bulletin				
4 p.m.	5 p.m.	6 p.m.	7 p.m.	Slow Code	Fast Code	Slow Code	Fast Code	Slow Code
5 p.m.	6 p.m.	7 p.m.	8 p.m.	Code Bulletin				
6 p.m.	7 p.m.	8 p.m.	9 p.m.	Digital Bulletin				
6:45 p.m.	7:45 p.m.	8:45 p.m.	9:45 p.m.	Voice Bulletin				
7 p.m.	8 p.m.	9 p.m.	10 p.m.	Fast Code	Slow Code	Fast Code	Slow Code	Fast Code
8 p.m.	9 p.m.	10 p.m.	11 p.m.	Code Bulletin				

CW is broadcast on the following MHz frequencies: 1.8025, 3.5815, 7.0475, 14.0475, 18.0975, 21.0675, 28.0675, and 147.555. W1AW schedule courtesy of *QST* magazine.

CODE COURSES ON CDs

Five words per minute is so slow, and so easy, that many ham radio applicants learn it completely in a single week! You can do it, too, by using my Morse code audio course.

Code courses personally recorded by me make code learning *fun.* They will train you to send and receive the International Morse code in just a few short weeks. They are narrated and parallel the instructions in this book. The CDs have code characters generated at a 15-wpm character rate, spaced out to a 5-wpm word rate. This is known as Farnsworth spacing.

Getting Started

The hardest part of learning the code is taking the first CD out of the case, putting it in your player, and pushing the play button! Try it, and you will be over your biggest hurdle. After that, the CDs will talk you through the code in no time at all.

The CDs make code learning *fun.* You'll hear how humor has been added to the learning process to keep your interest high. Since ham radio is a hobby, there's no reason we can't poke ourselves in the ribs and have a little fun learning the code as part of this hobby experience. Okay, you're still not convinced — you probably have already made up your mind that trying to learn the code will be the hardest part of being a ham. It will not. Give yourself a fair chance. Don't get discouraged. Have patience and remember these important reminders when practicing to learn the Morse code:

- Learn the code by sound. Don't stare at the tiny dots and dashes that we have here in the book — the dit and dah sounds on the air and with your practice keyer will ultimately create an instant letter at your fingertips and into the pencil.
- *Neve*r scribble down dots or dashes if you forget a letter. Just put a small dash on your paper for a missed letter. You can go back and figure out what the word is by the letters you did copy!
- Practice only with fast code characters; 15-wpm character speed, spaced down to 5-wpm speed, is ideal.
- Practice the code by writing it down whenever possible. This further trains your brain and hand to work together in a subconscious response to the sounds you hear. (Remember Pavlov and his dog "Spot"?)
- Practice only for 15 minutes at a time. The code course will tell you when to start and when to stop. Your brain and hand will lose that sharp edge once you go beyond 16 minutes of continuous code copy. You will learn much faster with five 15-minute practices per day than a one-hour marathon at night.
- Stay on course with the audio instructions. Learn the letters, numbers, punctuation marks, and operating signals in the order they are presented. My code teaching system parallels that of the American Radio Relay League, Boy Scouts of America, the Armed Forces, and has worked for thousands in actual classroom instruction.

It was no accident that Samuel Morse gave the single dit for the letter "E" which occurs most often in the English language. He determined the most used letters in the alphabet by counting letters in a printer's type case. He reasoned a printer would have more of the most commonly-used letters. It worked! With just the first lesson, you will be creating simple words and simple sentences with no previous background.

Table 6-2 shows the sequence of letters, punctuation marks, operating signals, and numbers covered in six lessons on the CDs recorded by me.

Table 6-2. Sequence of Lessons on Cassettes

• Lesson 1	E T M A N I S O $\overline{SK}$ Period
• Lesson 2	R U D C 5 Ø $\overline{AR}$ Question Mark
• Lesson 3	K P B G W F H $\overline{BT}$ Comma
• Lesson 4	Q L Y J X V Z $\overline{DN}$ 1 2 3 4 6 7 8 9
• Lesson 5	Random code with narrated answers
• Lesson 6	A typical 5-wpm code test

CODE KEY AND OSCILLATOR – HAM RECEIVER

All worldwide ham transceivers have provisions for a code key to be plugged in for both CW practice off the air as well as CW operating on the air. If you already own a worldwide set, chances are all you will need is a code key for some additional code-sending practice.

Read over your worldwide radio instruction manual where it talks about hooking up the code key. For code practice, read the notes about operating with a "side tone" but not actually going on the air. This "side tone" capability of most worldwide radios will eliminate your need for a separate code oscillator.

Code Key and Oscillator — Separate Unit

Many students may wish to simply buy a complete code key and oscillator set. They are available from local electronic outlets or through advertisements in the ham magazines.

Look again at the code key in *Figure 6-2a.* Note the terminals for the wires. Connect wires to these terminals and tighten the terminals so the wires won't come loose. The two wires will go either to a code oscillator set or to a plug that connects into your ham transceiver. Hook up the wires to the plug as described in your ham transceiver instruction book or the code oscillator set instruction book.

Mount the key firmly, as previously described, then adjust the gap between the contact points. With most new telegraph keys, you will need a pair of pliers to loosen the contact adjustment knob. It's located on the very end of your keyer. First loosen the lock nut, then screw down the adjustment until you get a gap no wider than the thickness of a business card. You want as little space as possible between the points. The contact points are located close to the sending plastic knob.

Now turn on your set or oscillator and listen. If your hear a constant tone, check that the right-hand movable shorting bar is not closed. If it is, swing it open. Adjust the spring tension adjustment screw so that you get a good "feel" each time you push down on the key knob. Adjust it tight enough to keep the contacts from closing while your fingers are resting on the key knob.

Pick up the key by the knob! This is the exact position your fingers should grasp the knob—one or two on top, and one or two on the side of it. Poking at the knob with one finger is unacceptable. Letting your fingers fly off the knob between dots and dashes (dits and dahs) also is not correct. As you are sending, you should be able to instantly pick up the whole key assembly to verify proper finger position.

Your arm and wrist should barely move as you send CW. All the action is in your hand — and it should be almost effortless. Give it a try, and look at *Figure6-2b* again to double-check your hand position.

Letting someone else use the key to send CW to you will also help you learn the code.

Morse Code Computer Software

The newest way to learn Morse code is through computer-aided instruction. There are many good PC programs on the market that not only teach you the characters, but build speed and allow you to take actual telegraphy examinations, which the computer constructs. Software programs also can be used to make digital CW sessions to play on your smart phone, too.

A big advantage of computer-aided Morse code learning is that you can easily customize the program to fit your own needs! You can select the sending speed, Farnsworth character-spacing speed, duration of transmission, number of characters in a random group, tone frequency — and more!

Some have built-in "weighting." That means the software will determine your weaknesses and automatically adjust future sending to give you more study on your problem characters! All Morse code software programs transmit the tone by keying the PC's internal speaker. Some generate a clearer audio tone through the use of external oscillators or internal computer sound cards.

Get my code course and start listening to my voice, and see how easy it is to master the dots and dashes. Continuously push yourself to the letters with more dots and dashes in them, and work those CDs regularly and keep your copy in your spiral-bound notebook.

I hope to hear your CW call on the worldwide bands soon!

Watch This! **https://www.youtube.com/playlist?list=PL1KAjn5rGhizk5f2_whBLJaOHj59OCl_3 (videos 202, 215 + 358)**

APPENDIX

U.S. VOLUNTEER EXAMINER COORDINATORS IN THE AMATEUR SERVICE

The five VECs listed with an * are those that offer remote examinations.

*W5YI-VEC
P.O. Box 200065
Arlington, TX 76006-0065
P: 800-669-9594
E: w5yi-vec@w5yi.org
Internet: W5YI.org

*Anchorage ARC VEC
P.O. Box 190192
Anchorage, AK 99519
P: (907) 632-3963
E: vec@kl7aa.org
Internet: KL7AA.net

*American Radio Relay League (ARRL)
225 Main Street
Newington, CT 06111-1494
P: 860-594-0300
F: 860-594-0339
E: VEC@arrl.org
Internet: ARRL.org

*Greater L.A. Amateur Radio Group
POB 500133
Palmdale, CA 93591
P: 661-264-1863
E: vec@glaarg.org
Internet: glaarg.org

*WCARS-VEC
417 IVY HILL Rd.
Weaverville, NC 28787
P: 828-658-0261
E: N4SET.wnc@Frontier.com
Internet wcars-vec.org

Central America CAVEC, Inc.
1204 Governors Dr SE
Huntsville, AL 35801
P: 256-509-5271
E: CAVEC.org@gmail.com
Internet:cavechamexam.com

Golden Empire Amateur Radio Society (GEARS)
P.O. Box 508
Chico, CA 95927-0508
P: 530-893-9211
E: myw6js@gmail.com
Internet:gearsw6rhc.com

Jefferson Amateur Radio Club
P.O. Box 73665
Metairie, LA 70033
P: 504-636-8809
E: w5gad.vec@w5gad.org
Internet: w5gad.org

Laurel Amateur Radio Club, Inc.
5287 W Belmont Rd
Tucson, AZ 85743
P: 520-664-5070
E: aa3of@arrl.net
Internet: larcmd.org

MRAC VEC, Inc. (Milwaukee)
2505 S. Calhoun Rd., #203
New Berlin, WI 53151
E: MRACVEC@gmail.com
Internet: www.mracvec.org

MO-KAN VEC Coordinator
228 Tennessee Road
Richmond, KS 66080-9174
P: 785-615-1097
E: bill.wo0e@gmail.com

Sandarc-VEC
5511 Maryland Ave
La Mesa, CA 91942-1519
P: 619-843-3747
E: n6nyx@arrl.net
Internet: sandarc.org

Sunnyvale VEC Amateur Radio Club, Inc.
P.O. Box 60307
Sunnyvale, CA 94088-0307
P: 408-255-9000
E: vec@amateur-radio.org
Internet:www.amateur-radio.org

W4VEC Volunteer Examiners Club of America
P. O. Box 41
Lexington, NC 27293-0041
P: 336-249-8734
E: raef@lexcominc.net
Internet: w4vec.org

THE W5YI RF SAFETY TABLES

(Developed by Fred Maia, W5YI, working in cooperation with the ARRL.)

There are two ways to determine whether your station's radio frequency signal radiation is within the MPE (Maximum Permissible Exposure) guidelines established by the FCC for ***"controlled"*** and ***"uncontrolled"*** environments. One way is direct ***"measurement"*** of the RF fields. The second way is through ***"prediction"*** using various antenna modeling, equations and calculation methods described in the FCC's *OET Bulletin 65* and *Supplement B.*

In general, most amateurs will not have access to the appropriate calibrated equipment to make precise field strength/power density measurements. The field-strength meters in common use by amateur operators are inexpensive, handheld field strength meters that do not provide the accuracy necessary for reliable measurements, especially when different frequencies may be encountered at a given measurement location. It is more practical for amateurs to determine their PEP output power at the antenna and then look up the required distances to the controlled/uncontrolled environments using the following tables, which were developed using the prediction equations supplied by the FCC.

The FCC has determined that radio operators and their families are in the "controlled" environment and your neighbors and passers-by are in the "uncontrolled" environment. The estimated minimum compliance distances are in meters from the transmitting antenna to either the occupational/controlled exposure environment ("Con") or the general population/uncontrolled exposure environment ("Unc") using typical antenna gains for the amateur service and assuming 100% duty cycle and maximum surface reflection. Therefore, these charts represent the worst case scenario. They do not take into consideration compliance distance reductions that would be caused by:

(1) Feed line losses, which reduce power output at the antenna especially at the VHF and higher frequency levels.

(2) Duty cycle caused by the emission type. The emission type factor accounts for the fact that, for some modulated emission types that have a non-constant envelope, the PEP can be considerably larger than the average power. Multiply the distances by 0.4 if you are using CW Morse telegraphy, and by 0.2 for two-way SSB (single sideband) voice. There is no reduction for FM.

(3) Duty cycle caused by on/off time or "time-averaging." The RF safety guidelines permit RF exposures to be averaged over certain periods of time with the average not to exceed the limit for continuous exposure. The averaging time for occupational/controlled exposures is 6 minutes, while the averaging time for general population/uncontrolled exposures is 30 minutes. For example, if the relevant time interval for time-averaging is 6 minutes, an amateur could be exposed to two times the applicable power density limit for three minutes as long as he or she were not exposed at all for the preceding or following three minutes.

A routine evaluation is also needed for vehicular mobile or handheld transceiver stations. Amateur Radio operators should be aware, however, of the potential for exposure to RF electromagnetic fields from these stations, and take measures (such as reducing transmitting power to the minimum necessary, positioning the radiating antenna as far from humans as practical, and limiting continuous transmitting time) to protect themselves and the occupants of their vehicles.

Amateur Radio operators should also be aware that the FCC radio-frequency safety regulations address exposure to people — and not the strength of the signal. Amateurs may exceed the Maximum Permissible Exposure (MPE) limits as long as no one is exposed to the radiation.

How to read the chart: If you are radiating 500 watts from your 10 meter dipole (about a 3 dB gain), there must be at least 4.5 meters (about 15 feet) between you (and your family) and the antenna — and a distance of 10 meters (about 33 feet) between the antenna and your neighbors.

Medium and High Frequency Amateur Bands

All distances are in meters

Freq. (MF/HF) (MHz/Band)	Antenna Gain (dBi)	Peak Envelope Power (watts) 100 watts		500 watts		1000 watts		1500 watts	
		Con.	Unc.	Con.	Unc.	Con.	Unc.	Con.	Unc.
2.0 (160m)	0	0.1	0.2	0.3	0.5	0.5	0.7	0.6	0.8
2.0 (160m)	3	0.2	0.3	0.5	0.7	0.6	1.06	0.8	1.2
4.0 (75/80m)	0	0.2	0.4	0.4	1.0	0.6	1.3	0.7	1.6
4.0 (75/80m)	3	0.3	0.6	0.6	1.3	0.9	1.9	1.0	2.3
7.3 (40m)	0	0.3	0.8	0.8	1.7	1.1	2.5	1.3	3.0
7.3 (40m)	3	0.5	1.1	1.1	2.5	1.6	3.5	1.9	4.2
7.3 (40m)	6	0.7	1.5	1.5	3.5	2.2	4.9	2.7	6.0
10.15 (30m)	0	0.5	1.1	1.1	2.4	1.5	3.4	1.9	4.2
10.15 (30m)	3	0.7	1.5	1.5	3.4	2.2	4.8	2.6	5.9
10.15 (30m)	6	1.0	2.2	2.2	4.8	3.0	6.8	3.7	8.3
14.35 (20m)	0	0.7	1.5	1.5	3.4	2.2	4.8	2.6	5.9
14.35 (20m)	3	1.0	2.2	2.2	4.8	3.0	6.8	3.7	8.4
14.35 (20m)	6	1.4	3.0	3.0	6.8	4.3	9.6	5.3	11.8
14.35 (20m)	9	1.9	4.3	4.3	9.6	6.1	13.6	7.5	16.7
18.168 (17m)	0	0.9	1.9	1.9	4.3	2.7	6.1	3.3	7.5
18.168 (17m)	3	1.2	2.7	2.7	6.1	3.9	8.6	4.7	10.6
18.168 (17m)	6	1.7	3.9	3.9	8.6	5.5	12.2	6.7	14.9
18.168 (17m)	9	2.4	5.4	5.4	12.2	7.7	17.2	9.4	21.1
21.145 (15m)	0	1.0	2.3	2.3	5.1	3.2	7.2	4.0	8.8
21.145 (15m)	3	1.4	3.2	3.2	7.2	4.6	10.2	5.6	12.5
21.145 (15m)	6	2.0	4.6	4.6	10.2	6.4	14.4	7.9	17.6
21.145 (15m)	9	2.9	6.4	6.4	14.4	9.1	20.3	11.1	24.9
24.99 (12m)	0	1.2	2.7	2.7	5.9	3.8	8.4	4.6	10.3
24.99 (12m)	3	1.7	3.8	3.8	8.4	5.3	11.9	6.5	14.5
24.99 (12m)	6	2.4	5.3	5.3	11.9	7.5	16.8	9.2	20.5
24.99 (12m)	9	3.4	7.5	7.5	16.8	10.6	23.7	13.0	29.0
29.7 (10m)	0	1.4	3.2	3.2	7.1	4.5	10.0	5.5	12.2
29.7 (10m)	3	2.0	4.5	4.5	10.0	6.3	14.1	7.7	17.3
29.7 (10m)	6	2.8	6.3	6.3	14.1	8.9	19.9	10.9	24.4
29.7 (10m)	9	4.0	8.9	8.9	19.9	12.6	28.2	15.4	34.5

VHF/UHF Amateur Bands

All distances are in meters

Freq. (MF/HF) (MHz/Band)	Antenna Gain (dBi)	Peak Envelope Power (watts) 50 watts		100 watts		500 watts		1000 watts	
		Con.	Unc.	Con.	Unc.	Con.	Unc.	Con.	Unc.
50 (6m)	0	1.0	2.3	1.4	3.2	3.2	7.1	4.5	10.1
50 (6m)	3	1.4	3.2	2.0	4.5	4.5	10.1	6.4	14.3
50 (6m)	6	2.0	4.5	2.8	6.4	6.4	14.2	9.0	20.1
50 (6m)	9	2.8	6.4	4.0	9.0	9.0	20.1	12.7	28.4
50 (6m)	12	4.0	9.0	5.7	12.7	12.7	28.4	18.0	40.2
50 (6m)	15	5.7	12.7	8.0	18.0	18.0	40.2	25.4	56.8
144 (2m)	0	1.0	2.3	1.4	3.2	3.2	7.1	4.5	10.1
144 (2m)	3	1.4	3.2	2.0	4.5	4.5	10.1	6.4	14.3
144 (2m)	6	2.0	4.5	2.8	6.4	6.4	14.2	9.0	20.1
144 (2m)	9	2.8	6.4	4.0	9.0	9.0	20.1	12.7	28.4
144 (2m)	12	4.0	9.0	5.7	12.7	12.7	28.4	18.0	40.2
144 (2m)	15	5.7	12.7	8.0	18.0	18.0	40.2	25.4	56.8
144 (2m)	20	10.1	22.6	14.3	32.0	32.0	71.4	45.1	101.0
222 (1.25m)	0	1.0	2.3	1.4	3.2	3.2	7.1	4.5	10.1
222 (1.25m)	3	1.4	3.2	2.0	4.5	4.5	10.1	6.4	14.3
222 (1.25m)	6	2.0	4.5	2.8	6.4	6.4	14.2	9.0	20.1
222 (1.25m)	9	2.8	6.4	4.0	9.0	9.0	20.1	12.7	28.4
222 (1.25m)	12	4.0	9.0	5.7	12.7	12.7	28.4	18.0	40.2
222 (1.25m)	15	5.7	12.7	8.0	18.0	18.0	40.2	25.4	56.8
450 (70cm)	0	0.8	1.8	1.2	2.6	2.6	5.8	3.7	8.2
450 (70cm)	3	1.2	2.6	1.6	3.7	3.7	8.2	5.2	11.6
450 (70cm)	6	1.6	3.7	2.3	5.2	5.2	11.6	7.4	16.4
450 (70cm)	9	2.3	5.2	3.3	7.3	7.3	16.4	10.4	23.2
450 (70cm)	12	3.3	7.3	4.6	10.4	10.4	23.2	14.7	32.8
902 (33cm)	0	0.6	1.3	0.8	1.8	1.8	4.1	2.6	5.8
902 (33cm)	3	0.8	1.8	1.2	2.6	2.6	5.8	3.7	8.2
902 (33cm)	6	1.2	2.6	1.6	3.7	3.7	8.2	5.2	11.6
902 (33cm)	9	1.6	3.7	2.3	5.2	5.2	11.6	7.3	16.4
902 (33cm)	12	2.3	5.2	3.3	7.3	7.3	16.4	10.4	23.2
1240 (23cm)	0	0.5	1.1	0.7	1.6	1.6	3.5	2.2	5.0
1240 (23cm)	3	0.7	1.6	1.0	2.2	2.2	5.0	3.1	7.0
1240 (23cm)	6	1.0	2.2	1.4	3.1	3.1	7.0	4.4	9.9
1240 (23cm)	9	1.4	3.1	2.0	4.4	4.4	9.9	6.3	14.0
1240 (23cm)	12	2.0	4.4	2.8	6.2	6.2	14.0	8.8	19.8

All distances are in meters. To convert from meters to feet multiply meters by 3.28. Distance indicated is shortest line-of-sight distance to point where MPE limit for appropriate exposure tier is predicted to occur.

AUTHORIZED FREQUENCY BANDS – AMATEUR SERVICE (for U.S. Amateur Stations operating from ITU-Region 2–North and South America)

Current License Class[1]		Technician	General		Extra Class
Grandfathered[2]	**Novice**	**Technician**	**General**	**Advanced**	**Extra Class**
2200			135.7-137.8 kHz/All	135.7-137.8 kHz/All	135.7-137.8 kHz/All
630			472-479 kHz/All	472-479 kHz/All	472-479 kHz/All
160			1800-2000 kHz/All	1800-2000 kHz/All	1800-2000 kHz/All
80/75	3525-3600 kHz/CW	3525-3600 kHz/CW	3525-3600 kHz/CW 3800-4000 kHz/Ph	3525-3600 kHz/CW 3700-4000 kHz/Ph	3500-4000 kHz/CW 3600-4000 kHz/Ph
60			5 channels	5 channels	5 channels
40	7025-7125 kHz/CW	7025-7125 kHz/CW	7025-7125 kHz/CW 7175-7300 kHz/Ph	7025-7125 kHz/CW 7125-7300 kHz/Ph	7000-7300 kHz/CW 7125-7300 kHz/Ph
30			10.1-10.15 MHz/CW	10.1-10.15 MHz/CW	10.1-10.15 MHz/CW
20			14.025-14.15 MHz/CW 14.225-14.35 MHz/Ph	14.025-14.15 MHz/CW 14.175-14.35 MHz/Ph	14.0-14.35 MHz/CW 14.15-14.35 MHz/Ph
17			18.068-18.11 MHz/CW 18.11-18.168 MHz/Ph	18.068-18.11 MHz/CW 18.11-18.168 MHz/Ph	18.068-18.11 MHz/CW 18.11-18.168 MHz/Ph
15	21.025-21.2 MHz/CW	21.025-21.2 MHz/CW	21.025-21.2 MHz/CW 21.275-21.45 MHz/Ph	21.025-21.2 MHz/CW 21.225-21.45 MHz/Ph	21.0-21.45 MHz/CW 21.2-21.45 MHz/Ph
12			24.89-24.99 MHz/CW 24.93-24.99 MHz/Ph	24.89-24.99 MHz/CW 24.93-24.99 MHz/Ph	24.89-24.99 MHz/CW 24.93-24.99 MHz/Ph
10	28.0-28.5 MHz/CW 28.3-28.5 MHz/Ph	28.0-28.5 MHz/CW 28.3-28.5 MHz/Ph	28.0-28.3 MHz/CW 28.3-29.7 MHz/Ph	28.0-28.3 MHz/CW 28.3-29.7 MHz/Ph	28.0-29.7 MHz/CW 28.3-29.7 MHz/Ph
6		50-54 MHz/CW 50.1-54 MHz/Ph	50-54 MHz/CW 50.1-54 MHz/Ph	50-54 MHz/CW 50.1-54 MHz/Ph	50-54 MHz/CW 50.1-54 MHz/Ph
2		144-148 MHz/CW 144.1-148 MHz/All	144-148 MHz/CW 144.1-148 MHz/Ph	144-148 MHz/CW 144.1-148 MHz/All	144-148 MHz/CW 144.1-148 MHz/All
1.25	222-225 MHz/All	222-225 MHz/All	222-225 MHz/All	222-225 MHz/All	222-225 MHz/All
0.70		420-450 MHz/All	420-450 MHz/All	420-450 MHz/All	420-450 MHz/All
0.33		902-928 MHz/All	902-928 MHz/All	902-928 MHz/All	902-928 MHz/All
0.23	1270-1295 MHz/All	1240-1300 MHz/All	1240-1300 MHz/All	1240-1300 MHz/All	1240-1300 MHz/All

METERS

[1] Effective 4-15-00 [2] Prior to 4-15-00

COMMON CW ABBREVIATIONS

AA	All after
AB	All before
ABT	About
ADR	Address
AGN	Again
ANT	Antenna
BCI	Broadcast interference
BK	Break; break me; break in
BN	All between; been
B4	Before
C	Yes
CFM	Confirm; I confirm
CK	Check
CL	I am closing my station; call
CLD-CLG	Called; calling
CUD	Could
CUL	See you later
CUM	Come
CW	Continuous Wave
DLD-DLVD	Delivered
DX	Distance
FB	Fine business; excellent
GA	Go ahead (or resume sending)
GB	Good-by
GBA	Give better address
GE	Good evening
GG	Going
GM	Good morning
GN	Good night
GND	Ground
GUD	Good
HI	The telegraphic laugh; high
HR	Here; hear
HV	Have
HW	How
LID	A poor operator
MILS	Milliamperes
MSG	Message; prefix to radiogram
N	No
ND	Nothing doing
NIL	Nothing; I have nothing for you
NR	Number
NW	Now; I resume transmission
OB	Old boy
OM	Old man
OP-OPR	Operator
OT	Old timer; old top
PBL	Preable
PSE-PLS	Please
PWR	Power
PX	Press
R	Received as transmitted; are
RCD	Received
REF	Refer to; referring to; reference
RPT	Repeat; I repeat
SED	Said
SEZ	Says
SIG	Signature; signal
SKED	Schedule
SRI	Sorry
SVC	Service; prefix to service message
TFC	Traffic
TMW	Tomorrow
TNX	Thanks
TU	Thank you
TVI	Television interference
TXT	Text
UR-URS	Your; you're; yours
VFO-	Variable-frequency oscillator
VY	Very
WA	Word after
WB	Word before
WD-WDS	Word; words
WKD-WKG	Worked; working
WL	Well; will
WUD	Would
WX	Weather
XMTR	Transmitter
XTAL	Crystal
XYL	Wife
YL	Young lady
73	Best regards
88	Love and kisses

QUESTION POOL SYLLABUS

The syllabus used by the NCVEC Question Pool Committee to develop the question pool is included here as an aid in studying the subelements and topic groups. Reviewing the syllabus will give you and understanding of how the question pool is used to develop the Element 2 written theory examination. Remember, one question will be taken from each topic group within each subelement to create your exam.

2022-2026 Element 2 (Technician Class) Syllabus

SUBELEMENT T1 – COMMISSION'S RULES - [6 Exam Questions - 6 Groups] {Modified} 67 Questions

T1A - Purpose and permissible use of the Amateur Radio Service; Operator/primary station license grant; Meanings of basic terms used in FCC rules; Interference; RACES rules; Phonetics; Frequency Coordinator {Modified}

T1B - Frequency allocations; Emission modes; Spectrum sharing; Transmissions near band edges; Contacting the International Space Station; Power output {Modified}

T1C - Licensing: classes, sequential and vanity call sign systems, places where the Amateur Radio Service is regulated by the FCC, name and address on FCC license database, term, renewal, grace period, maintaining mailing address; International communications {Modified}

T1D - Authorized and prohibited transmissions: communications with other countries, music, exchange of information with other services, indecent language, compensation for operating, retransmission of other amateur signals, encryption, sale of equipment, unidentified transmissions, one-way transmission {Modified}

T1E - Control operator: eligibility, designating, privileges, duties, location, required; Control point; Control types: automatic, remote {Modified}

T1F - Station identification; Repeaters; Third party communications; Club stations; FCC inspection {Modified}

SUBELEMENT T2 - OPERATING PROCEDURES - [3 Exam Questions - 3 Groups] 36 Questions

T2A - Station operation: choosing an operating frequency, calling another station, test transmissions; Band plans: calling frequencies, repeater offsets {Modified}

T2B – VHF/UHF operating practices: FM repeater, simplex, reverse splits; Access tones: CTCSS, DTMF; DMR operation; Resolving operational problems; Q signals {Modified}

T2C – Public service: emergency operations, applicability of FCC rules, RACES and ARES, net and traffic procedures, operating restrictions during emergencies, use of phonetics in message handling {Modified}

SUBELEMENT T3 – RADIO WAVE PROPAGATION – [3 Exam Questions - 3 Groups] {Modified} 34 Questions

T3A - Radio wave characteristics: how a radio signal travels, fading, multipath, polarization, wavelength vs absorption; Antenna orientation {Modified}

T3B - Electromagnetic wave properties: wavelength vs frequency, nature and velocity of electromagnetic waves, relationship of wavelength and frequency; Electromagnetic spectrum definitions: UHF, VHF, HF {Modified}

T3C - Propagation modes: sporadic E, meteor scatter, auroral propagation, tropospheric ducting; F region skip; Line of sight and radio horizon {Modified}

SUBELEMENT T4 – AMATEUR RADIO PRACTICES – [2 Exam Questions - 2 Groups] {Modified} 24 Questions

T4A – Station setup: connecting a microphone, a power source, a computer, digital equipment, an SWR meter; bonding; Mobile radio installation {Modified}

T4B - Operating controls: frequency tuning, use of filters, squelch function, AGC, memory channels, noise blanker, microphone gain, receiver incremental tuning (RIT), bandwidth selection, digital transceiver configuration {Modified}

SUBELEMENT T5 – ELECTRICAL PRINCIPLES – [4 Exam Questions - 4 Groups] {Modified} 52 Questions

T5A – Current and voltage: terminology and units, conductors and insulators, alternating and direct current {Modified}

T5B - Math for electronics: conversion of electrical units, decibels {Modified}

T5C – Capacitance and inductance terminology and units; Radio frequency definition and units; Impedance definition and units; Calculating power {Modified}

T5D – Ohm's Law; Series and parallel circuits {Modified}

SUBELEMENT T6 – ELECTRONIC AND ELECTRICAL COMPONENTS – [4 Exam Questions - 4 Groups] {Modified} 47 Questions

T6A - Fixed and variable resistors; Capacitors; Inductors; Fuses; Switches; Batteries {Modified}

T6B – Semiconductors: basic principles and applications of solid state devices, diodes and transistors {Modified}

T6C - Circuit diagrams: use of schematics, basic structure; Schematic symbols of basic components {Modified}

T6D - Component functions: rectifiers, relays, voltage regulators, meters, indicators, integrated circuits, transformers; Resonant circuit; Shielding {Modified}

SUBELEMENT T7 – PRACTICAL CIRCUITS – [4 Exam Questions - 4 Groups] {Modified} 44 Questions

T7A – Station equipment: receivers, transceivers, transmitter amplifiers, receive amplifiers, transverters; Basic radio circuit concepts and terminology: sensitivity, selectivity, mixers, oscillators, PTT, modulation {Modified}

T7B – Symptoms, causes, and cures of common transmitter and receiver problems: overload and overdrive, distortion, interference and consumer electronics, RF feedback {Modified}

T7C – Antenna and transmission line measurements and troubleshooting: measuring SWR, effects of high SWR, causes of feed line failures; Basic coaxial cable characteristics; Use of dummy loads when testing {Modified}

T7D – Using basic test instruments: voltmeter, ammeter, and ohmmeter; Soldering {Modified}

SUBELEMENT T8 – SIGNALS AND EMISSIONS – [4 Exam Questions - 4 Groups] {Modified} 48 Questions

T8A – Basic characteristics of FM and SSB; Bandwidth of various modulation modes: CW, SSB, FM, fast-scan TV; Choice of emission type: selection of USB vs LSB, use of SSB for weak signal work, use of FM for VHF packet and repeaters {Modified}

T8B - Amateur satellite operation: Doppler shift, basic orbits, operating protocols, modulation mode selection, transmitter power considerations, telemetry and telecommand, satellite tracking programs, beacons, uplink and downlink mode definitions, spin fading, definition of "LEO", setting uplink power {Modified}

T8C – Operating activities: radio direction finding, contests, linking over the internet, exchanging grid locators {Modified}

T8D – Non-voice and digital communications: image signals and definition of NTSC, CW, packet radio, PSK, APRS, error detection and correction, amateur radio networking, Digital Mobile Radio, WSJT modes, Broadband-Hamnet {Modified}

SUBELEMENT T9 – ANTENNAS AND FEED LINES - [2 Exam Questions - 2 Groups] 24 Questions

T9A – Antennas: vertical and horizontal polarization, concept of antenna gain, definition and types of beam antennas, antenna loading, common portable and mobile antennas, relationships between resonant length and frequency, dipole pattern {Modified}

T9B – Feed lines: types, attenuation vs frequency, selecting; SWR concepts; Antenna tuners (couplers); RF Connectors: selecting, weather protection

SUBELEMENT T0 – SAFETY – [3 Exam Questions - 3 Groups] {Modified} 36 Questions

T0A – Power circuits and hazards: hazardous voltages, fuses and circuit breakers, grounding, electrical code compliance; Lightning protection; Battery safety {Modified}

T0B – Antenna safety: tower safety and grounding, installing antennas, antenna supports {Modified}

T0C - RF hazards: radiation exposure, proximity to antennas, recognized safe power levels, radiation types, duty cycle {Modified}

2022-2026 ELEMENT 2 Q&A CROSS REFERENCE

The following cross reference list presents all 411 question numbers included in the 2022-2026 Question Pool in numerical order, followed by the page number on which each question begins in this book. This will allow you to locate specific questions by the question number.

T1 – Commission's Rules

Question	Page
T1A01	33
T1A02	33
T1A03	39
T1A04	35
T1A05	34
T1A06	97
T1A07	101
T1A08	82
T1A09	84
T1A10	86
T1A11	51
T1B01	99
T1B02	102
T1B03	61
T1B04	62
T1B05	62
T1B06	96
T1B07	107
T1B08	63
T1B09	63
T1B10	97
T1B11	60
T1B12	61
T1C01	34
T1C02	38
T1C03	42
T1C04	54
T1C05	39
T1C06	43
T1C07	54
T1C08	35
T1C09	35
T1C10	34
T1C11	36

Question	Page
T1D01	41
T1D02	52
T1D03	52
T1D04	53
T1D05	53
T1D06	51
T1D07	80
T1D08	49
T1D09	52
T1D10	52
T1D11	38
T1E01	45
T1E02	46
T1E03	46
T1E04	46
T1E05	48
T1E06	47
T1E07	46
T1E08	48
T1E09	49
T1E10	48
T1E11	49
T1F01	54
T1F02	40
T1F03	37
T1F04	40
T1F05	84
T1F06	40
T1F07	43
T1F08	42
T1F09	79
T1F10	50
T1F11	39

T2 – Operating Procedures

Question	Page
T2A01	81
T2A02	62
T2A03	81
T2A04	73
T2A05	73
T2A06	72
T2A07	80
T2A08	72
T2A09	82
T2A10	63
T2A11	71
T2A12	73
T2B01	80
T2B02	81
T2B03	70
T2B04	81
T2B05	130
T2B06	113
T2B07	126
T2B08	75
T2B09	71
T2B10	75
T2B11	77
T2B12	126
T2B13	69
T2C01	86
T2C02	87
T2C03	89
T2C04	86
T2C05	88
T2C06	85
T2C07	85
T2C08	89
T2C09	87
T2C10	88
T2C11	89

T3 - Propagation

Question	Page
T3A01	75
T3A02	92
T3A03	172
T3A04	173
T3A05	169
T3A06	75
T3A07	92
T3A08	99
T3A09	98
T3A10	111
T3A11	96
T3A12	91
T3B01	55
T3B02	99
T3B03	55
T3B04	56
T3B05	58
T3B06	59
T3B07	58
T3B08	60
T3B09	60
T3B10	59
T3B11	57
T3C01	92
T3C02	96
T3C03	94
T3C04	98
T3C05	93
T3C06	93
T3C07	94
T3C08	94
T3C09	97
T3C10	98
T3C11	91

T4 – Am Radio Practices

Question	Page
T4A01	139
T4A02	179
T4A03	138
T4A04	110
T4A05	180
T4A06	108
T4A07	109
T4A08	191
T4A09	136
T4A10	109
T4A11	137
T4A12	108
T4B01	129
T4B02	69
T4B03	70
T4B04	65
T4B05	130
T4B06	121
T4B07	126
T4B08	122
T4B09	126
T4B10	122
T4B11	127
T4B12	73

T5 – Electrical Principles

Question	Page
T5A01	137
T5A02	149
T5A03	137
T5A04	140
T5A05	135
T5A06	57
T5A07	137
T5A08	141
T5A09	137
T5A10	149

Glossary

Amateur communication: Noncommercial radio communication by or among amateur stations solely with a personal aim and without personal or business interest.

Amateur operator/primary station license: An instrument of authorization issued by the FCC comprised of a station license, and also incorporating an operator license indicating the class of privileges.

Amateur operator: A person holding a valid license to operate an amateur station issued by the FCC. Amateur operators are frequently referred to as ham operators.

Amateur Radio services: The amateur service, the amateur-satellite service, and the radio amateur civil emergency service.

Amateur-satellite service: A radiocommunication service using stations on Earth satellites for the same purpose as those of the amateur service.

Amateur service: A radiocommunication service for the purpose of self-training, intercommunication and technical investigations carried out by amateurs; that is, duly authorized persons interested in radio technique solely with a personal aim and without pecuniary interest.

Amateur station: A station licensed in the amateur service embracing necessary apparatus at a particular location used for amateur communication.

AMSAT: Radio Amateur Satellite Corporation, a nonprofit scientific organization. (P.O. Box #27, Washington, DC 20044)

ANSI: American National Standards Institute. A non-government organization that develops recommended standards for a variety of applications.

APRS: Automatic Position Radio System, which takes GPS (Global Positioning System) information and translates it into an automatic packet of digital information.

ARES: Amateur Radio Emergency Service — the emergency division of the American Radio Relay League. Also see RACES

ARRL: American Radio Relay League, national organization of U.S. Amateur Radio operators. (225 Main Street, Newington, CT 06111)

Audio Frequency (AF): The range of frequencies that can be heard by the human ear, generally 20 hertz to 20 kilohertz.

Automatic control: The use of devices and procedures for station control without the control operator being present at the control point when the station is transmitting.

Automatic Volume Control (AVC): A circuit that continually maintains a constant audio output volume in spite of deviations in input signal strength.

Beam or Yagi antenna: An antenna array that receives or transmits RF energy in a particular direction. Usually rotatable.

Block diagram: A simplified outline of an electronic system where circuits or components are shown as boxes.

Broadcasting: Information or programming transmitted by radio means intended for the general public.

Bulletin No. 65: The Office of Engineering & Technology bulletin that provides specified safety guidelines for human exposure to radiofrequency (RF) radiation.

Business communications: Any transmission or communication the purpose of which is to facilitate the regular business or commercial affairs of any party. Business communications are prohibited in the amateur service.

Call Book: A published list of all licensed amateur operators available in North American and Foreign editions.

Call sign: The FCC systematically assigns each amateur station its primary call sign.

Certificate of Successful Completion of Examination (CSCE): A certificate providing examination credit for 365 days. Both written and code credit can be authorized.

Coaxial cable, Coax: A concentric, two-conductor cable in which one conductor surrounds the other, separated by an insulator.

Controlled Environment: Involves people who are aware of and who can exercise control over radiofrequency exposure. Controlled exposure limits apply to both occupational workers and Amateur Radio operators and their immediate households.

Control operator: An amateur operator designated by the licensee of an amateur station to be responsible for the station transmissions.

Coordinated repeater station: An amateur repeater station for which the transmitting and receiving frequencies have been recommended by the recognized repeater coordinator.

Coordinated Universal Time (UTC): (Also Greenwich Mean Time, UCT or Zulu time.) The time at the zero-degree (0°) Meridian which passes through Greenwich, England. A universal time among all amateur operators.

Crystal: A quartz or similar material which has been ground to produce natural vibrations of a specific frequency. Quartz crystals produce a high degree of frequency stability in radio transmitters.

CW: See Morse code.

Dipole antenna: The most common wire antenna. Length is equal to one-half of the wavelength. Fed by coaxial cable.

Dummy antenna: A device or resistor which serves as a transmitter's antenna without radiating radio waves. Generally used to tune up a radio transmitter.

Duplexer: A device that allows a single antenna to be simultaneously used for both reception and transmission.

Duty cycle: As applies to RF safety, the percentage of time that a transmitter is "on" versus "off" in a 6- or 30-minute time period.

Effective Radiated Power (ERP): The product of the transmitter (peak envelope) power, expressed in watts, delivered to the antenna, and the relative gain of an antenna over that of a half-wave dipole antenna.

Electromagnetic radiation: The propagation of radiant energy, including infrared, visible light, ultraviolet, radiofrequency, gamma and X-rays, through space and matter.

Emergency communication: Any amateur communication directly relating to the immediate safety of life of individuals or the immediate protection of property.

Examination Element: The written theory exam or CW test required for various classes of FCC Amateur Radio licenses. Technician must pass Element 2 written theory; General must pass Element 3 written theory plus Element 1 CW; Extra must pass Element 4 written theory.

Far Field: The electromagnetic field located at a great distance from a transmitting antenna. The far field begins at a distance that depends on many factors, including the wavelength and the size of the antenna. Radio signals are normally received in the far field.

FCC Form 605: The FCC application form used to apply for a new amateur operator/primary station license or to renew or modify an existing license.

Federal Communications Commission (FCC): A board of five Commissioners, appointed by the President, having the power to regulate wire and radio telecommunications in the U.S.

Feedline: A system of conductors that connects an antenna to a receiver or transmitter.

Field Day: Annual activity sponsored by the ARRL to demonstrate emergency preparedness of amateur operators.

Field strength: A measure of the intensity of an electric or magnetic field. Electric fields are measured in volts per meter; magnetic fields in amperes per meter.

Filter: A device used to block or reduce alternating currents or signals at certain frequencies while allowing others to pass unimpeded.

Frequency: The number of cycles of alternating current in one second.

Frequency coordinator: An individual or organization which recommends frequencies and other operating and/or technical parameters for amateur repeater operation in order to avoid or minimize potential interferences.

Frequency Modulation (FM): A method of varying a radio carrier wave by causing its frequency to vary in accordance with the information to be conveyed.

Frequency privileges: The transmitting frequency bands available to the various classes of amateur operators. The various Class privileges are listed in Part 97.301 of the FCC rules.

Ground: A connection, accidental or intentional, between a device or circuit and the earth or some common body and the earth or some common body serving as the earth.

Ground wave: A radio wave that is propagated near or at the earth's surface.

Handi-Ham system: Amateur organization dedicated to assisting handicapped amateur operators. (3915 Golden Valley Road, Golden Valley, MN 55422)

Harmful interference: Interference which seriously degrades, obstructs or repeatedly interrupts the operation of a radio communication service.

Harmonic: A radio wave that is a multiple of the fundamental frequency. The second harmonic is twice the fundamental frequency, the third harmonic, three times, etc.

Hertz: One complete alternating cycle per second. Named after Heinrich R. Hertz, a German physicist. The number of hertz is the frequency of the audio or radio wave.

High Frequency (HF): The band of frequencies that lie between 3 and 30 Megahertz. It is from these frequencies that radio waves are returned to earth from the ionosphere.

High-Pass filter: A device that allows passage of high frequency signals but attenuates the lower frequencies. When installed on a television set, a high-pass filter allows TV frequencies to pass while blocking lower-frequency amateur signals.

Inverse Square Law: The physical principle by which power density decreases as you get further away from a transmitting antenna. RF power density decreases by the inverse square of the distance.

Ionization: The process of adding or stripping away electrons from atoms or molecules. Ionization occurs when substances are heated at high temperatures or exposed to high voltages. It can lead to significant genetic damage in biological tissue.

Ionosphere: Outer limits of atmosphere from which HF amateur communications signals are returned to earth.

IRC: International Reply Coupon, a method of prepaying postage for a foreign amateur's QSL card.

Jamming: The intentional, malicious interference with another radio signal.

Key clicks, Chirps: Defective keying of a telegraphy signal sounding like tapping or high varying pitches.

Linear amplifier: A device that accurately reproduces a radio wave in magnified form.

Long wire: A horizontal wire antenna that is one wavelength or longer in length.

Low-Pass filter: Device connected to worldwide transmitters that inhibits passage of higher frequencies that cause television interference but does not affect amateur transmissions.

Machine: A ham slang word for an automatic repeater station.

Malicious interference: See jamming.

MARS: The Military Affiliate Radio System. An organization that coordinates the activities of amateur communications with military radio communications.

Maximum authorized transmitting power: Amateur stations must use no more than the maximum transmitter power necessary to carry out the desired communications. The maximum P.E.P. output power levels authorized Novices are 200 watts in the 80-, 40-, 15- and 10-meter bands, 25 watts in the 222-MHz band, and 5 watts in the 1270-MHz bands.

Maximum Permissible Exposure (MPE): The maximum amount of electric and magnetic RF energy to which a person may safely be exposed.

Maximum usable frequency (MUF): The highest frequency that will be returned to earth from the ionosphere.

Medium frequency (MF): The band of frequencies that lies between 300 and 3,000 kHz (3 MHz).

Microwave: Electromagnetic waves with a frequency of 300 MHz to 300 GHz. Microwaves can cause heating of biological tissue.

Mobile operation: Radio communications conducted while in motion or during halts at unspecified locations.

Mode: Type of transmission such as voice, teletype, code, television, facsimile.

Modulate: To vary the amplitude, frequency, or phase of a radiofrequency wave in accordance with the information to be conveyed.

Morse code: The International Morse code, A1A emission. Interrupted continuous wave communications conducted using a dot-dash code for letters, numbers and operating procedure signs.

Near Field: The electromagnetic field located in the immediate vicinity of the antenna. Energy in the near field depends on the size of the antenna, its wavelength and transmission power.

Nonionizing radiation: Electromagnetic waves, or fields, which do not have the capability to alter the molecular structure of substances. RF energy is nonionizing radiation.

Novice operator: An FCC licensed, entry-level amateur operator in the amateur service.

Occupational exposure: See controlled environment.

OET: Office of Engineering & Technology, a branch of the FCC that has developed the guidelines for radiofrequency (RF) safety.

Ohm's law: The basic electrical law explaining the relationship between voltage, current and resistance. The current (I) in a circuit is equal to the voltage (E) divided by the resistance (R), or I = E/R.

OSCAR: "Orbiting Satellite Carrying Amateur Radio." A series of satellites designed and built by amateur operators of several nations.

Oscillator: A device for generating oscillations or vibrations of an audio or radiofrequency signal.

Packet radio: A digital method of communicating computer-to-computer. A terminal-node controller makes up the packet of data and directs it to another packet station.

Peak Envelope Power (PEP): 1. The power during one radiofrequency cycle at the crest of the modulation envelope, taken under normal operating conditions. 2. The maximum power that can be obtained from a transmitter.

Phone patch: Interconnection of amateur radio to the public switched telephone network, and operated by the control operator of the station.

Power density: A measure of the strength of an electromagnetic field at a distance from its source. Usually expressed in milliwatts per square centimeter (mW/cm2). Far-field power density decreases according to the Law of Inverse Squares.

Power supply: A device or circuit that provides the appropriate voltage and current to another device or circuit.

Propagation: The travel of electromagnetic waves or sound waves through a medium.

Public exposure: See "uncontrolled" environment.

Q-signals: International three-letter abbreviations beginning with the letter Q used primarily to convey information using the Morse code.

QSL Bureau: An office that bulk processes QSL (radio confirmation) cards for (or from) foreign amateur operators as a postage-saving mechanism.

RACES (Radio Amateur Civil Emergency Service): A radio service using amateur stations for civil defense communications during periods of local, regional, or national emergencies.

Radiation: Electromagnetic energy, such as radio waves, traveling forth into space from a transmitter.

Radiofrequency (RF): The range of frequencies over 20 kilohertz that can be propagated through space.

Radiofrequency (RF) radiation: Electromagnetic fields or waves having a frequency between 3 kHz and 300 GHz.

Radiofrequency spectrum: The eight electromagnetic bands ranked according to their frequency and wavelength. Specifically, the very-low, low, medium, high, very-high, ultra-high, super-high, and extremely-high frequency bands.

Radio wave: A combination of electric and magnetic fields varying at a radiofrequency and traveling through space at the speed of light.

Repeater operation: Automatic amateur stations that retransmit the signals of other amateur stations.

Routine RF radiation evaluation: The process of determining if the RF energy from a transmitter exceeds the Maximum Permissible Exposure (MPE) limits in a controlled or uncontrolled environment.

RST Report: A telegraphy signal report system of Readability, Strength and Tone.

S-meter: A voltmeter calibrated from 0 to 9 that indicates the relative signal strength of an incoming signal at a radio receiver.

Selectivity: The ability of a circuit (or radio receiver) to separate the desired signal from those not wanted.

Sensitivity: The ability of a circuit (or radio receiver) to detect a specified input signal.

Short circuit: An unintended, low-resistance connection across a voltage source resulting in high current and possible damage.

Shortwave: The high frequencies that lie between 3 and 30 Megahertz that are propagated long distances.

Single-Sideband (SSB): A method of radio transmission in which the RF carrier and one of the sidebands is suppressed and all of the information is carried in the one remaining sideband.

Skip wave, Skip zone: A radio wave reflected back to earth. The distance between the radio transmitter and the site of a radio wave's return to earth.

Sky-wave: A radio wave that is refracted back to earth. Sometimes called an ionospheric wave.

Specific Absorption Rate (SAR): The time rate at which radiofrequency energy is absorbed into the human body.

Spectrum: A series of radiated energies arranged in order of wavelength. The radio spectrum extends from 20 kilohertz upward.

Spurious Emissions: Unwanted radiofrequency signals emitted from a transmitter that sometimes cause interference.

Station license, location: No transmitting station shall be operated in the amateur service without being licensed by the FCC. Each amateur station shall have one land location, the address of which appears in the station license.

Sunspot Cycle: An 11-year cycle of solar disturbances which greatly affects radio wave propagation.

Technician operator: An Amateur Radio operator who has successfully passed Element 2.

Technician-Plus: An amateur operator who has passed a 5-wpm code test in addition to Technician Class requirements.

Telegraphy: Communications transmission and reception using CW, International Morse code.

Telephony: Communications transmission and reception in the voice mode.

Telecommunications: The electrical conversion, switching, transmission and control of audio video and data signals by wire or radio.

Temporary operating authority: Authority to operate your amateur station while awaiting arrival of an upgraded license.

Terrestrial station location: Any location of a radio station on the surface of the earth including the sea.

Thermal effects: As applies to RF radiation, biological tissue damage resulting because of the body's inability to cope with or dissipate excessive heat.

Third-party traffic: Amateur communication by or under the supervision of the control operator at an amateur station to another amateur station on behalf of others.

Time-averaging: As applies to RF safety, the amount of electromagnetic radiation over a given time. The premise of time-averaging is that the human body can tolerate the thermal load caused by high, localized RF exposures for short periods of time.

Transceiver: A combination radio transmitter and receiver.

Transition region: Area where power density decreases inversely with distance from the antenna.

Transmatch: An antenna tuner used to match the impedance of the transmitter output to the transmission line of an antenna.

Transmitter: Equipment used to generate radio waves. Most commonly, this radio carrier signal is amplitude varied or frequency varied (modulated) with information and radiated into space.

Transmitter power: The average peak envelope power (output) present at the antenna terminals of the transmitter. The term "transmitted" includes any external radio-frequency power amplifier which may be used.

Ultra High Frequency (UHF): Ultra high frequency radio waves that are in the range of 300 to 3,000 MHz.

Uncontrolled environment: Applies to those persons who have no control over their exposure to RF energy in the environment. Residences adjacent to ham radio installations are considered to be in an "uncontrolled" environment.

Upper Sideband (USB): The proper operating mode for sideband transmissions made in the new Novice 10-meter voice band. Amateurs generally operate USB at 20 meters and higher frequencies; lower sideband (LSB) at 40 meters and lower frequencies.

Very High Frequency (VHF): Very high frequency radio waves that are in the range of 30 to 300 MHz.

Volunteer Examiner: An amateur operator of at least a General Class level who prepares and administers amateur operator license examinations.

Volunteer Examiner Coordinator (VEC): A member of an organization which has entered into an agreement with the FCC to coordinate the efforts of volunteer examiners in preparing and administering examinations for amateur operator licenses.

Index

FCC Commercial Radio License Study Materials

General Radiotelephone Operator's License

GROL+RADAR
by Gordon West, WB6NOA with Technical Editor Eric P. Nichols, KL7AJ. Get your FCC commercial radio licenses and add valuable credentials to your resume! *GROL+RADAR* includes the new FCC Element 1 question pool for the Marine Radio Operator Permit (MROP), the Element 3 pool for the General Radiotelephone Operator License (GROL), and the Element 8 pool for the RADAR Endorsement. Many employers require these licenses for jobs in marine, aero, safety, and municipal positions. Gordo and his team have written clear explanations for all the Q&A to make studying for these exams educational and fun. If you're an Extra Class ham, many of the technical/math questions will look familiar to you. Fully-illustrated to aid your learning.
GROL $49.95

GROL+RADAR Book & Software Package Enhance your learning experience using our practice exam software along with the *GROL+RADAR* book. Software includes answer explanations from the book – when you select a wrong answer, the explanation from the book appears to reinforce your learning. **GRSP $79.95**

Learn More With Our 'Basic' Series

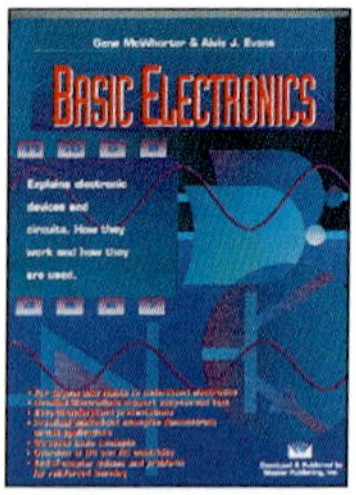

Basic Electronics
by Gene McWhorter & Alvis Evans is for anyone who wants to understand the fundamentals of electronics. Explains electronic devices and circuits – how they work and are used. Contains detailed illustrations that support easy-to-read text. Practical, worked-out examples demonstrate circuit applications.
BELC $19.95

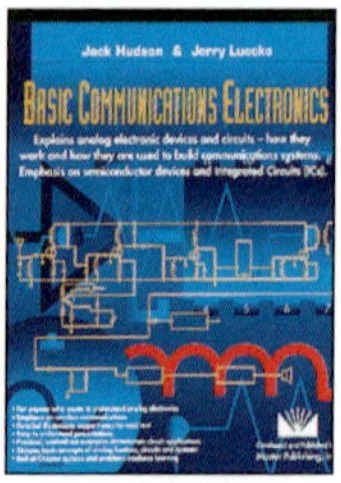

Basic Communications Electronics
by Jack Hudson, W9MU, & Jerry Luecke, KB5TZY. Explains analog electronic devices and circuits – how they work and how they are used to build communications systems. Emphasis on semiconductor devices and integrated circuits (ICs). For anyone who wants to understand analog electronics and wireless communications. **BCOM $19.95**

Basic Digital Electronics
by Alvis Evans. Explains digital system functions and how circuits are used to build them. Discusses the functions required to design digital systems, the circuits used to make decisions, code conversions, data selections, interfacing and storage, and the circuits that keep all operations in time and under control.
BDIG $19.95

The Worldwide Listening Guide

By John Figliozzi. Modeled on his popular *Worldwide Shortwave Listening Guide*, this new book explains radio listening in all today's formats – "live," on-demand, WiFi, podcast, terrestrial, satellite, internet, digital and, of course, analog, AM, FM and SW. Book includes a comprehensive programming guide to what can be heard how, when and where. Spiral-bound to open in a flat, easy-to-use format. **WWLG $29.95**

Mail Ordering Information

If you would like to order any of our products by mail, send your check or money order to:

W5YI Group
POB 200065
Arlington, TX 76006-0065

SHIPPING: Please add $8.00 for the first item, and $1.00 for each additional item. Items are shipped via USPS Priority Mail.